PUBLIC WATER SUPPLY DISTRIBUTION SYSTEMS: ASSESSING AND REDUCING RISKS

FIRST REPORT

Committee on Public Water Supply Distribution Systems: Assessing and Reducing Risks

Water Science and Technology Board

Division on Earth and Life Studies

NATIONAL RESEARCH COUNCIL
OF THE NATIONAL ACADEMIES

THE NATIONAL ACADEMIES PRESS
Washington, D.C.
www.nap.edu

THE NATIONAL ACADEMIES PRESS   500 Fifth Street, N.W.   Washington, DC 20001

NOTICE: The project that is the subject of this report was approved by the Governing Board of the National Research Council, whose members are drawn from the councils of the National Academy of Sciences, the National Academy of Engineering, and the Institute of Medicine. The members of the committee responsible for the report were chosen for their special competences and with regard for appropriate balance.

Support for this project was provided by EPA Contact No. 68-C-03-081. Any opinions, findings, conclusions, or recommendations expressed in this publication are those of the author(s) and do not necessarily reflect the views of the organizations or agencies that provided support for the project.

Additional copies of this report are available from the National Academies Press, 500 Fifth Street, N.W., Lockbox 285, Washington, DC 20055; (800) 624-6242 or (202) 334-3313 (in the Washington metropolitan area); Internet, *http://www.nap.edu*.

International Standard Book Number 0-309-09628-6

Copyright 2005 by the National Academy of Sciences. All rights reserved.

Printed in the United States of America.

THE NATIONAL ACADEMIES
Advisers to the Nation on Science, Engineering, and Medicine

The **National Academy of Sciences** is a private, nonprofit, self-perpetuating society of distinguished scholars engaged in scientific and engineering research, dedicated to the furtherance of science and technology and to their use for the general welfare. Upon the authority of the charter granted to it by the Congress in 1863, the Academy has a mandate that requires it to advise the federal government on scientific and technical matters. Dr. Bruce M. Alberts is president of the National Academy of Sciences.

The **National Academy of Engineering** was established in 1964, under the charter of the National Academy of Sciences, as a parallel organization of outstanding engineers. It is autonomous in its administration and in the selection of its members, sharing with the National Academy of Sciences the responsibility for advising the federal government. The National Academy of Engineering also sponsors engineering programs aimed at meeting national needs, encourages education and research, and recognizes the superior achievement of engineers. Dr. Wm. A. Wulf is president of the National Academy of Engineering.

The **Institute of Medicine** was established in 1970 by the National Academy of Sciences to secure the services of eminent members of appropriate professions in the examination of policy matters pertaining to the health of the public. The Institute acts under the responsibility given to the National Academy of Sciences by its congressional charter to be an adviser to the federal government and, upon its own initiative, to identify issues of medical care, research, and education. Dr. Harvey V. Fineberg is president of the Institute of Medicine.

The **National Research Council** was organized by the National Academy of Sciences in 1916 to associate the broad community of science and technology with the Academy's purposes of furthering knowledge and advising the federal government. Functioning in accordance with general policies determined by the Academy, the Council has become the principal operating agency of both the National Academy of Sciences and the National Academy of Engineering in providing services to the government, the public, and the scientific and engineering communities. The Council is administered jointly by both Academies and the Institute of Medicine. Dr. Bruce M. Alberts and Dr. Wm. A. Wulf are chair and vice-chair, respectively, of the National Research Council.

www.national-academies.org

COMMITTEE ON PUBLIC WATER SUPPLY DISTRIBUTION SYSTEMS: ASSESSING AND REDUCING RISKS

Vernon L. Snoeyink, *Chair*, University of Illinois, Urbana-Champaign
Charles N. Haas, *Vice-Chair*, Drexel University, Philadelphia, Pennsylvania
Paul F. Boulos, MWH Soft, Broomfield, Colorado
Gary A. Burlingame, Philadelphia Water Department, Philadelphia, Pennsylvania
Anne K. Camper, Montana State University, Bozeman
Robert N. Clark, Shaw Environmental and Infrastructure, Inc., Cincinnati, Ohio
Marc A. Edwards, Virginia Polytechnic and State University, Blacksburg
Mark W. LeChevallier, American Water Corporation, Voorhees, New Jersey
L. D. McMullen, Des Moines Water Works, Des Moines, Iowa
Christine L. Moe, Emory University, Atlanta, Georgia
Eva C. Nieminski, Utah Department of Environmental Quality, Salt Lake City
Charlotte D. Smith, Charlotte Smith and Associates, Inc., Orinda, California
David P. Spath, California Department of Health Services, Sacramento
Gary A. Toranzos, University of Puerto Rico, San Juan
Richard L. Valentine, University of Iowa, Iowa City

National Research Council Staff

Laura J. Ehlers, Study Director
Ellen A. De Guzman, Research Associate

WATER SCIENCE AND TECHNOLOGY BOARD

R. RHODES TRUSSELL, *Chair,* Trussell Technologies, Inc., Pasadena, California
MARY JO BAEDECKER, U.S. Geological Survey (Retired), Vienna, Virginia
GREGORY B. BAECHER, University of Maryland, College Park
JOAN G. EHRENFELD, Rutgers University, New Brunswick, New Jersey
DARA ENTEKHABI, Massachusetts Institute of Technology, Cambridge, Massachusetts
GERALD E. GALLOWAY, Titan Corporation, Reston, Virginia
PETER GLEICK, Pacific Institute for Studies in Development, Environment, and Security, Oakland, California
CHARLES N. HAAS, Drexel University, Philadelphia, Pennsylvania
KAI N. LEE, Williams College, Williamstown, Massachusetts
CHRISTINE L. MOE, Emory University, Atlanta, Georgia
ROBERT PERCIASEPE, National Audubon Society, New York, New York
JERALD L. SCHNOOR, University of Iowa, Iowa City
LEONARD SHABMAN, Resources for the Future, Washington, DC
KARL K. TUREKIAN, Yale University, New Haven, Connecticut
HAME M. WATT, Independent Consultant, Washington, DC
CLAIRE WELTY, University of Maryland, Baltimore County
JAMES L. WESCOAT, JR., University of Illinois, Urbana-Champaign

Staff

STEPHEN D. PARKER, Director
LAURA J. EHLERS, Senior Staff Officer
JEFFREY W. JACOBS, Senior Staff Officer
WILLIAM S. LOGAN, Senior Staff Officer
LAUREN E. ALEXANDER, Staff Officer
STEPHANIE E. JOHNSON, Staff Officer
M. JEANNE AQUILINO, Financial and Administrative Associate
ELLEN A. DE GUZMAN, Research Associate
PATRICIA JONES KERSHAW, Study/Research Associate
ANITA A. HALL, Administrative Assistant
DOROTHY K. WEIR, Senior Project Assistant

Preface

The distribution system is a critical component of every drinking water utility. Its primary function is to provide the required water quantity and quality at suitable pressure, and failure to do so is a serious system deficiency. Water quality may degrade during water distribution because of the way water is treated or not treated before it is distributed, chemical and biological reactions that take place in the water during distribution, reactions between the water and distribution system materials, and contamination from external sources that occurs because of main breaks, leaks coupled with hydraulic transients, improperly maintained storage facilities, and other factors. Special problems are posed by the utility's need to maintain suitable water quality at the consumers tap, and the quality changes that occur in consumers' plumbing, which is not owned or controlled by the utility. The primary driving force for managing and regulating distribution systems is protecting the health of the consumer, but certainly factors that cause water of poor aesthetic quality to be delivered to the tap or that increase the cost of delivering water are also important. Our nation's distribution systems are aging and becoming more vulnerable to main breaks and leaks, possibly because they are underground and out of sight, and thus it is easy to delay distribution system investment when budgets are considered. There is an urgent need for new research that will enable cost-effective treatment for distribution and design, construction, and management of the distribution system for protection of public health and minimization of water quality degradation.

This study of the Water Science and Technology Board of the National Research Council (NRC) was undertaken at the request of the U.S. Environmental Protection Agency (EPA). The purpose of this report is to:

- Identify trends relevant to the deterioration of drinking water quality in water supply distribution systems, using available information, and
- Identify and prioritize issues of greatest concern for distribution systems based on review of published material.

This first report was requested by the EPA, as it considers revisions to the Total Coliform Rule in 2005. It will be followed in about 18 months by a more comprehensive final report that evaluates different approaches to characterization of public health risks posed by water-quality deteriorating events, identifies and evaluates the effectiveness of relevant existing codes and regulations, and identifies general actions, strategies, performance measures, and policies that could be considered by water utilities and other stakeholders to reduce the risks posed by water-quality deteriorating events or conditions. Advances in detection, monitoring and modeling, analytical methods, information needs and technologies, research and development opportunities, and communication strategies that will enable the water supply industry and other stakeholders to further reduce risks associated with public water supply distribution systems will also be addressed.

This report has been reviewed by individuals chosen for their diverse perspectives and technical expertise, in accordance with the procedures approved by the NRC's Report Review Committee. The purpose of this independent review is to provide candid and critical comments that will assist the authors and the NRC in making the published report as sound as possible and to ensure that the report meets institutional standards of objectivity, evidence, and responsiveness to the study charge. The reviews and draft manuscripts remain confidential to protect the integrity of the deliberative process. We thank the following individuals for their participation in the review of this report: Gunther F. Craun, Craun and Associates, Virginia; Jerry Ongerth, East Bay Municipal Utility District, California; Jerald L. Schnoor, The University of Iowa, Iowa; R. Rhodes Trussell, Trussell Technologies, Inc., Califonia; and Jack Wang, Louisville Water Company, Kentucky.

Although the reviewers listed above have provided many constructive comments and suggestions, they were not asked to endorse the conclusions or recommendations nor did they see the final draft of the report before its release. The review of this report was overseen by Edward Bouwer of the Johns Hopkins University. Appointed by the NRC, he was responsible for making certain that an independent examination of this report was carried out in accordance with institutional procedures and that all review comments were carefully considered. Responsibility for the final content of this report rests entirely with the authoring committee.

Vernon Snoeyink, *Chair*

Contents

1 INTRODUCTION 1

2 TRENDS RELEVANT TO THE DETERIORATION OF DRINKING WATER
IN DISTRIBUTION SYSTEMS 4
    Distribution Pipe Age and Replacement Rates 4
    Waterborne Disease Outbreaks 6
    Changes in the United States Population 7
    Use of Bottled Water and Household Water Treatment Devices 8

3 HIGHEST PRIORITY ISSUES 10
    Cross Connections and Backflow 11
    New and Repaired Water Mains 12
    Finished Water Storage 14
    Additional Issues of Concern 15

4 MEDIUM PRIORITY ISSUES 19
    Biofilm Growth 19
    Loss of Residual via Water Age and Nitrification 20
    Low Pressure Transients and Intrusion 22

5 LOWER PRIORITY ISSUES 24
    Other Effects of Water Age 25
    Other Effects of Nitrification 25
    Permeation 26
    Leaching 27
    Additional Issues of Concern 28

6 SUMMARY 30

REFERENCES 35

APPENDIX A 43
    Committee Biographical Information

1
Introduction

The distribution systems of public drinking water supplies include the pipes and other conveyances that connect treatment plants to consumers' taps. They span almost 1 billion miles in the United States (Kirmeyer et al., 1994) and include an estimated 154,000 finished water storage facilities (AWWA, 2003). Public water supplies serve 273 million residential and commercial customers, although the vast majority (93 percent) of systems serves less than 10,000 people (EPA, 2004). As the U.S. population grows and communities expand, 13,200 miles of new pipes are installed each year (Kirmeyer et al., 1994).

Distribution systems constitute a significant management challenge from both an operational and public health standpoint. Furthermore, they represent the vast majority of physical infrastructure for water supplies, such that their repair and replacement represent an enormous financial liability. The U. S. Environmental Protection Agency (EPA) estimates the 20-year water transmission and distribution needs of the country to be $83.2 billion, with storage facility infrastructure needs estimated at $18.4 billion (EPA, 1999).

Most federal water quality regulations pertaining to drinking water, such as Maximum Contaminant Levels (MCLs) and treatment technique requirements for microbial and chemical contaminants, are applied before or at the point where water enters the distribution system. The major rules that specifically target water quality within the distribution system are the Lead and Copper Rule (LCR), the Surface Water Treatment Rule (SWTR), which addresses the minimum required detectable disinfectant residual and the maximum allowed heterotrophic bacterial plate count, and the Total Coliform Rule. In addition, the Disinfectants/Disinfection By-Products Rule (D/DBPR) addresses the maximum disinfectant residual and concentration of disinfection byproducts like total trihalomethanes and haloacetic acids allowed in distribution systems. Of all these rules, the Total Coliform Rule (TCR) of 1989 explicitly addresses microbial water quality in the distribution system. The TCR applies to all public water supplies, both groundwater and surface water, and requires (among other things) an MCL of less than 5 percent of water samples testing positive for total coliforms in any month for systems serving more than 33,000, and that there be no more than one positive sample per month for systems serving less than 33,000 (Guilaran, 2004). Sampling of distribution systems varies widely, from as many as hundreds of samples per month to one sample per year, depending on the size of the system. This and other information gathered since the rule was first drafted suggest that the TCR is limited in its ability to ensure public health protection from microbial contamination of distribution systems. Indeed, some epidemiological and outbreak investigations conducted in the last five years suggest that a substantial proportion of waterborne disease outbreaks, both microbial and chemical, is attributable to problems within distribution systems (Craun and Calderon, 2001; Blackburn et al., 2004). Distribution system deficiencies were pinpointed as the cause of 57 reported community outbreaks from 1991 to 1998 (EPA, 2002b). There is no evidence that the current regulatory program has resulted in a diminution in the proportion of outbreaks attributable to distribution system related factors.

In 2000, EPA drew attention to distribution systems in making recommendations for the Microbial/Disinfection By-products Rule (M/DBPR) by agreeing to evaluate available data and research on aspects of distribution systems that may create risks to public health. Furthermore, in 2003 EPA committed to revising the TCR—not only to update the provisions about the frequency and location of monitoring, follow-up monitoring after total coliform positive samples, and the basis of the MCL, but also to address the broader issue of whether the TCR could be revised to encompass "distribution system integrity." That is, EPA is exploring the possibility of revising the TCR to provide a comprehensive approach for addressing water quality in the distribution system environment. To aid in this process, EPA requested the input of the National Academies' Water Science and Technology Board, which was asked to conduct a study of water quality issues associated with public water supply distribution systems and their potential risks to consumers.

The expert committee formed to conduct the study will consider, but not be limited to, specific aspects of distribution systems such as cross connections and backflow, intrusion caused by pressure transients, nitrification, permeation and leaching, repair and replacement of water mains, aging infrastructure, and microbial growth. The committee's statement of task is to:

1—Identify trends relevant to the deterioration of drinking water in water supply distribution systems, as background and based on available information.

2—Identify and prioritize issues of greatest concern for distribution systems based on review of published material.

3—Focusing on the highest priority issues as revealed by task #2, (a) evaluate different approaches for characterization of public health risks posed by water-quality deteriorating events or conditions that may occur in public water supply distribution systems; and (b) identify and evaluate the effectiveness of relevant existing codes and regulations and identify general actions, strategies, performance measures, and policies that could be considered by water utilities and other stakeholders to reduce the risks posed by water-quality deteriorating events or conditions. Case studies, either at state or utility level, where distribution system control programs (e.g., Hazard Analysis and Critical Control Point System, cross connection control, etc.) have been successfully designed and implemented will be identified and recommendations will be presented in their context.

Introduction

4—Identify advances in detection, monitoring and modeling, analytical methods, information needs and technologies, research and development opportunities, and communication strategies that will enable the water supply industry and other stakeholders to further reduce risks associated with public water supply distribution systems.

This first report relates the committee's progress on Tasks 1 and 2—that is, trends relevant to the deterioration of distribution system water quality and the issues that the committee thinks are the highest priorities for consideration during TCR revision to encompass distribution system integrity. Conclusions and recommendations related to distribution system issues that EPA may want to take into consideration are sprinkled throughout the text, and a short summary of the committee's prioritization is given at the end.

2
Trends Relevant to the Deterioration of Drinking Water in Distribution Systems

In the past two decades, a number of changes have occurred that may affect the quality of drinking water in distribution systems, consumer exposure to tap water, and the consequent risks of exposure. This section discusses trends in pipe age in water distribution systems and pipe replacement rates, waterborne disease outbreaks, host susceptibility in the U.S. population, consumer use of bottled water, and installation of home water treatment devices. This is not a comprehensive list of all the factors that may affect water quality and health risks from distributions systems. Furthermore, for many of these factors, there are limited data on recent trends such that additional research is needed to better understand current practices.

DISTRIBUTION PIPE AGE AND REPLACEMENT RATES

There is a large range in the type and age of the pipes that make up American water distribution systems, depending on the population and economic booms of the previous century. For many cities, the periods of greatest population growth and urban expansion were during the late 1800s, around World War I, during the 1920s, and post-World War II. The water pipes installed during these growth periods differ in their manufacture, materials, and life span. The oldest cast iron pipes from the late 19th century are typically described as having an average useful lifespan of about 120 years because of the pipe wall thickness (AWWA, 2001; AWWSC, 2002). In the 1920s the manufacture of iron pipes changed to improve pipe strength, but the changes also produced a thinner wall. These pipes have an average life of about 100 years. Pipe manufacturing continued to evolve in the 1950s and 1960s with the introduction of ductile iron pipe that is stronger than cast iron and more resistant to corrosion. Polyvinyl chloride (PVC) pipes were introduced in the 1970s and high-density polyethylene in the 1990s. Both of these are very resistant to corrosion but they do not have the strength of ductile iron. Post-World War II pipes tend to have an average life of 75 years (AWWA, 2001; AWWSC, 2002). Approximately 20 percent of the pipe in place in North America is lined with asbestos or cement. Furthermore, the overwhelming majority of ductile iron pipe is mortar-lined and about 40 percent of cast iron pipe in place is mortar-lined. These facts may be of great importance where the life of pipe is concerned, as linings are meant to prevent corrosion and increase pipe longevity.

In the 20th century, most of the water systems and distribution pipes were relatively new and well within their expected lifespan. However, a recent report by the American Water Works Association (AWWA, 2001) and a white paper by the American Water Works Service Company, Inc. (AWWSC, 2002) point out that these different types of pipes, installed during different time periods, will all be reaching the end of their expected life spans in the next 30 years. Analysis of main breaks at one large Midwestern water utility that kept careful records of distribution system management documented a sharp increase in the annual number of main breaks from 1970 (approximately 250 breaks per year) to 1989 (approximately 2,200 breaks per year) (AWWSC, 2002). Thus, the water industry is entering an era where it must make substantial investments in pipe repair and pipe replacement. An EPA report on water infrastructure needs (EPA, 2002c) predicted that transmission and distribution replacement rates will need to be around 0.3 percent per year in 2005 and will rise to 2.0 percent per year by 2040 in order to adequately maintain the water infrastructure (see Figure 1). Cost estimates for drinking water infrastructure range from $4.2 to $6.3 billion per year (AWWSC, 2002). The trends of aging pipe and increasing numbers of main breaks are of concern because of the potential relationship between waterborne disease outbreaks and main breaks (see the subsequent section on New and Repaired Water Mains).

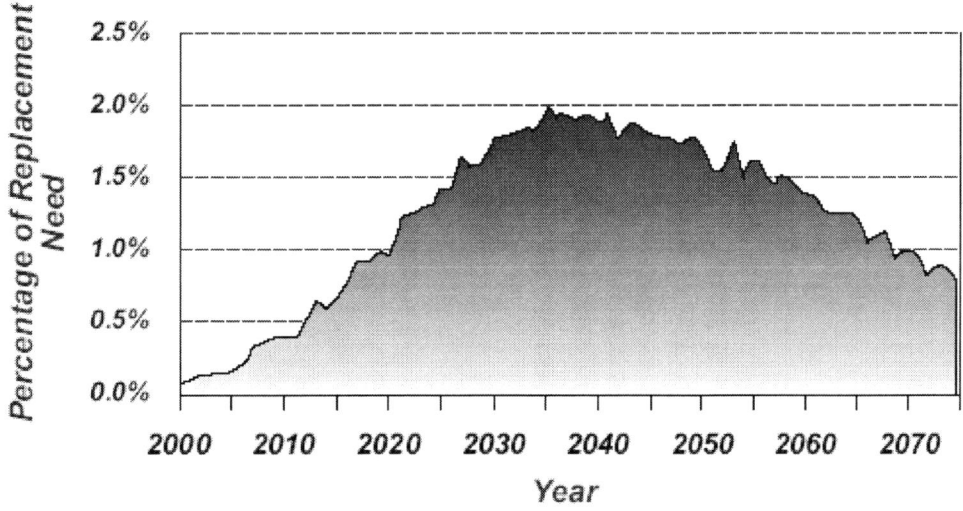

FIGURE 1 Projected annual replacement needs for transmission lines and distribution mains, 2000–2075. SOURCE: EPA (2002c).

WATERBORNE DISEASE OUTBREAKS

A voluntary, passive surveillance system for waterborne disease outbreaks in the U.S. has been maintained by the Centers for Disease Control and Prevention (CDC) in collaboration with EPA since 1971. Summary reports from this surveillance system are published every two years and describe the number of outbreaks, where they occurred, the etiologic agents, type of water systems involved, and factors that contributed to the outbreak. While the current waterborne disease surveillance summary states that the data are useful "for identifying major deficiencies in providing safe drinking water" (Blackburn et al., 2004), caution in the interpretation of these data is important, in that the proportion of outbreaks reported may vary with time, location, and the size of the water supply. With this caveat in mind, analyses of the data from this surveillance system indicate that the total number of reported waterborne disease outbreaks has decreased since 1980. However, the proportion of waterborne disease outbreaks associated with problems in the distribution system is increasing (see Figure 2). Craun and Calderon (2001) examined causes of reported waterborne outbreaks from 1971–1998 and noted that in community water systems, 30 percent of 294 outbreaks were associated with distribution system deficiencies. From 1999 to 2002, there have been 18 reported outbreaks in community water systems, and nine (50 percent) of these were related to problems in the water distribution system (Lee et al., 2002, Blackburn et al., 2004). The decrease in numbers of waterborne disease outbreaks per year is important and probably attributable

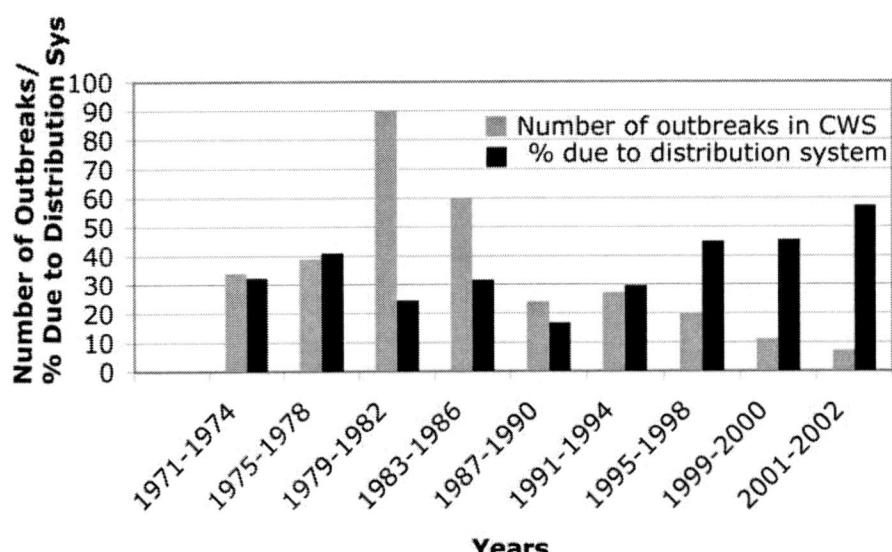

FIGURE 2 Waterborne disease outbreaks in community water systems (CWS) associated with distribution system deficiencies. Note that the majority of the reported outbreaks have been in small community systems and that the absolute numbers of outbreaks have decreased since 1982. SOURCE: Data extracted from Craun and Calderon (2001) and MMWR summary reports on waterborne disease surveillance (Lee et al., 2002 and Blackburn et al., 2004).

to improved water treatment practices and SWTR compliance that reduced the risk from waterborne protozoa (Pierson et al., 2001; Blackburn et al., 2004). The increase in the percentage of outbreaks attributable to the distribution system is probably also due to this factor (i.e., the SWTR); regulations for distribution systems have not been as extensive (other than the Lead and Copper Rule).

Most reported outbreaks associated with distribution systems occur in community water systems because of their greater size and complexity, but there have been a number of outbreaks associated with noncommunity water systems that have been attributed to deficiencies in the distribution system. Craun and Calderon (2001) reported that most distribution system-related outbreaks were linked to cross-connections and backsiphonage and the rest were attributed to main breaks or repair and contamination of municipal water storage tanks. The magnitude and severity of outbreaks associated with distribution systems vary, with an average of about 200 illnesses per outbreak (Craun and Calderon, 2001) and a total of 13 deaths. These outbreaks have been associated with chemical (copper, chlordane, ethylene glycol and others) and microbial contaminants, including enteric protozoa (*Giardia, Cyclospora*), enteric bacteria (*Salmonella, Shigella, Campylobacter*, and *E. coli* O157:H7) and enteric viruses (noroviruses and Hepatitis A virus).

It should be noted that endemic waterborne infection and illness may be associated with contaminants entering the distribution system. If low levels of contaminants enter the system and affect small numbers of persons, it might not be recognized and investigated as an outbreak. Indeed, it has been acknowledged that a fairly sizable number of cases of cryptosporidiosis could be occurring in a large city such as New York City without detection of a possible outbreak (NRC, 1999, page 249). Thus, not only are all waterborne disease outbreaks not detected, even those that are detected and reported will not address possible endemic illness risks. A number of sources show that endemic risks can be greater than epidemics, for example, Frost et al., (1996) and Payment and Hunter (2001). The CDC and EPA have recently completed a series of epidemiologic studies designed to assess the magnitude of endemic waterborne illness associated with consumption of municipal drinking water; a joint report on the results of these studies is forthcoming (Blackburn et al., 2004).

CHANGES IN THE UNITED STATES POPULATION

Another cause for concern regarding the risks of waterborne disease transmission is increasing host susceptibility to infection and disease in the U.S. population. Due to weaker immune systems, older Americans are at increased risk for morbidity and mortality from a number of infectious diseases, including influenza, pneumonia, and enteric diseases (Gerba et al., 1996). Decreased gastric acid secretion in the elderly may also result in increased susceptibility to infection from acid-sensitive enteric organisms (Morris and Potter, 1997). The U.S. population older than 74 years of age has increased from 13.1 million in 1990 to 16.6 million in 2000. The most rapid growth during this decade occurred in the size of the oldest age groups with a 38 percent increase in the population greater than 85 years of age. In 2003, more than 12 percent of the total U.S. population was 65 and older, and this proportion will increase dramatically between 2010 and 2030 as the "baby boomers" start turning 65 in 2011. By 2030, nearly 20 percent of the total U.S. population will be over 65 years of age, and the population over 85 years of age will have grown rapidly (Older Americans, 2004).

The numbers of immunocompromised persons in the U.S. due to disease and immunosuppressive therapy is also increasing. Of particular note are growing numbers of AIDS patients, cancer survivers, and organ transplant patients. The number of new AIDS cases reported in the U.S. has increased more than five-fold from 8,131 in 1985 to 44,232 in 2003. Because of more effective treatments, AIDS patients are living longer but are still at increased risk of enteric infections. Cancer patients and transplant patients often require immunosuppressive therapy that puts them at greater risk of infection during the course of their treatment. The CDC estimates that the number of persons living with cancer more than tripled from 3.0 million in 1971 (1.5 percent of the U.S. population) to 9.8 million in 2001 (3.5 percent of the population) (CDC, 2004). The number of organ transplants performed each year in the U.S. has almost doubled from 12,619 in 1988 to more than 22,554 in 2004 (Organ Procurement and Transplantation Network. Available on-line at *http://www.optn.org/latestData/rptData.asp*. Accessed February 6, 2005).

USE OF BOTTLED WATER AND HOUSEHOLD WATER TREATMENT DEVICES

There has been a dramatic increase in the proportion of the U.S. population that drinks bottled water or uses some type of water treatment device in their homes. The Natural Resources Defense Council (NRDC) reported in 1999 that more than half of all Americans drink bottled water and about one third of the population regularly drink bottled water. The sales of bottled water tripled between 1986 and 1997 and reached about $4 billion per year (NRDC, 1999). The International Bottled Water Association reported a 10.1 percent growth in sales between 1997 and 1998 (Available on-line at *http://www.bottledwater.org/public/pressrel.htm*. Accessed March 16, 2005). The cost of bottled water ranges from 240 to over 10,000 times more per gallon than tap water, and yet bottled water use is not limited to high-income households. One study reported that "Black, Asian, and Hispanic households are more likely than whites to use bottled water" despite lower household incomes (Mogelonsky, quoted in NRDC, 1999). The use of home water treatment devices has also risen steadily from 32 percent in 1997 to 38 percent in 1999 to 41 percent in 2001 (WQA, 2001). As discussed in a subsequent section, these devices can support the regrowth of microbes, such that there use is not necessarily correlated with a decrease in contaminant exposure.

Several consumer surveys and studies have attempted to determine the driving forces behind these trends and have reported that perceived safety and health, taste of tap water, and concern about some contaminants are the most frequently reported reasons people drink bottled water instead of tap water (NRDC, 1999; Anadu and Harding, 2000;

WQA, 2001; Mackey et al., 2003). Although these trends are occurring, the health implications of these trends are unknown.

***

Taken together, the trends data suggest that water distribution system infrastructure in the U.S. is deteriorating and that health risks associated with distribution system water quality may be increasing. Although the proportion of the U.S. population that may be more susceptible to waterborne disease is growing, fewer Americans are drinking tap water. These trends need to be investigated to determine if they are important factors that should be taken into account when developing a distribution system rule.

3
Highest Priority Issues

The second major task of the committee was to identify the highest priority issues for consideration during TCR revision to encompass distribution system integrity. The issues considered for prioritization stem directly from nine white papers created during a series of expert and stakeholder workshops convened by EPA and others from 2000 to 2003. The nine white papers focused on the following events or conditions that can bring about water quality degradation in public water supply distribution systems:

1. Cross-Connections and Backflow (EPA, 2002b)
2. Intrusion of Contaminants from Pressure Transients (LeChevallier et al., 2002)
3. Nitrification (AWWA and EES, Inc., 2002e)
4. Permeation and Leaching (AWWA and EES, Inc., 2002a)
5. Microbial Growth and Biofilms (EPA, 2002d)
6. New or Repaired Water Mains (AWWA and EES, Inc., 2002e)
7. Finished Water Storage Facilities (AWWA and EES, Inc., 2002c)
8. Water Age (AWWA and EES, Inc., 2002b)
9. Deteriorating Buried Infrastructure (AWWSC, 2002)

In addition to these papers, the committee considered the summary of the Distribution System Exposure Assessment Workshop (ICF Consulting, Inc., 2004), held in Washington, DC in March, 2004, which attempted to collate all of the information gathered in the previous workshops. Additional white papers are currently being written on the following topics, but were not available to the committee in time to be considered for this first report:

- Indicators of Drinking Water Quality
- Evaluation of Hazard Analysis and Critical Control Points
- Causes of Total Coliform Positives and Contamination Events
- Inorganic Contaminant Accumulation
- Distribution System Inventory and Condition Assessment.

Some qualitative outcomes of the many workshops, as communicated by EPA officials, are that there are demonstrated adverse health effects and large potential exposure resulting from distribution system contamination. The stakeholder and industry experts who attended the workshops agreed on the need to evaluate and prioritize potential health risk.

The approach to prioritization taken by the committee was based on a careful assessment of the issues presented in the nine white papers, critical evaluation of other materials, and on the committee's assessment of the health importance of the various events. Given limited data on the specific causes of waterborne disease outbreaks, the best professional judgment of the committee was used to assess the magnitude of the health problem associated with an event, including how often the event occurs and how much contamination results when an event occurs. In addition to prioritizing the issues presented in the nine white papers, the committee also considered whether any significant issues had been overlooked by EPA when the white papers were written.

It should be noted that EPA had a difficult task in developing these white papers. A water distribution system is a complex engineering and ecological system wherein multiple adverse changes may result from the same or similar underlying causes. For example, water that has a long residence time in the system (high water age) may also have the potential to lose disinfectant residual and undergo biological nitrification. Considering any of these occurrences in the absence of the others oversimplifies the nature of the problem. However, the committee decided to follow the structure of the EPA white papers in preparing this report, with the recognition that overlaps and difficult-to-separate phenomena exist.

Of the issues presented in the nine white papers, cross connections and backflow, new or repaired water mains, and finished water storage facilities were judged by the committee to be of the highest importance based on their associated potential health risks. In addition, there are two other issues that should also be accorded high priority: premise plumbing and distribution system operator training.

CROSS CONNECTIONS AND BACKFLOW

Points in a plumbing system where non-potable water comes into contact with the potable water supply are called cross connections. A backflow event occurs when non-potable water flows into the drinking water supply through a cross connection, either because of low distribution system pressure (termed backsiphonage) or because of pressure on the non-potable water caused by pumpage or other factors (termed backpressure). Backflow incidents have long been recognized as significant contributors to waterborne disease. From 1981 to 1998, the CDC documented 57 waterborne outbreaks related to cross-connections, resulting in 9,734 detected and reported illnesses (Craun and Calderon, 2001). EPA compiled a total of 459 incidents resulting in 12,093 illnesses from backflow events from 1970 to 2001 (EPA, 2002b). For the period 1981 to 1998, EPA found that only 97 of 309 incidents were reported to public health authorities, demonstrating that the magnitude of the public health concern due to cross-connections is underreported. The situation may be of even greater concern because incidents involving domestic plumbing are even less recognized. In a study of 188 households, the University of Southern California's Foundation for Cross-Connection Control and Hydraulic Research reported that

9.6 percent of the homes had a direct cross connection that constituted a health hazard, and more than 95 percent had either direct or indirect cross connections (USC, 2002). Cross-connections are also of great concern where a potable system is in close proximity to a reclaimed water system (such as in dual distribution systems like that of the Irvine Ranch water district). A direct cross connection is a permanent physical interconnection between potable and non-potable water sources whereas an indirect cross connection has the potential for an interconnection (e.g., a janitor's utility sink without a vacuum breaker on the hose bib). In most cases, the extent of the health problem caused by cross connections in homes is unknown—knowledge that could be obtained through epidemiological studies.

Although most state and primacy agencies require that utilities have a cross connection control program, the elements of such programs, their implementation, and oversight vary widely. Because of inconsistent application of these programs, cross connections and backflow events remain a significant potential cause of waterborne disease. Proven technologies and procedures are available to mitigate the impact of cross connections on potable water quality. State plumbing codes define the type of plumbing materials that are approved for use, including cross connection control devices, but whether these codes are adhered to is questionable. Regulatory options that could be considered include requiring inspections for household cross connections at the time of home sale. Furthermore, training programs such as those offered by the New England Water Works Association to train and certify backflow device installers and testers have been successful in gaining support from the plumbing community and in developing local plumbing codes that require cross connection control. Given the availability of effective technologies for preventing cross connections, opportunity exists for substantial reductions in public health risk through the implementation of more effective cross connection control programs by primacy and state agencies. At the current time, it is unknown how effective various state programs are in actually preventing cross connections—an issue that is also ripe for further investigation.

Because of the long history of recognized health risk posed by cross connections and backflow, the clear epidemiological and surveillance data, and the proven technologies to prevent cross connections, cross connection and backflow events are ranked by the committee as the highest priority. Efforts to provide implementation of a more uniform national cross connection program would have clear public health benefits.

NEW AND REPAIRED WATER MAINS

This section focuses on contamination arising from the exposure of distribution system water and pipe interior, appurtenances, and related materials to microbial and chemical contaminants in the external environment (1) during water main failures and breaks and (2) due to human activities to install new, rehabilitate old, or repair broken mains and appurtenances. When a pipe break or failure occurs, there is immediate potential for external contamination from soil, groundwater, or surface runoff (see Kirmeyer et al., 2001) to enter the distribution system or come into contact with the pipe interior in the area of the failure. Furthermore, the storage, installation, rehabilitation, and repair of water mains and appurtenances provide an opportunity for microbial and chemical contamination of materials that come into direct contact with drinking water. Pierson et al. (2001) confirmed the possibility of such events by surveying distribution system inspectors and field crews.

This section does not address contamination from the external environment that enters through cracks or leaks in pipe, pipe joints, or appurtenances (even though these can exist undetected for long periods of time), as these are covered under the intrusion section. Furthermore, periodic changes to the operation of the distribution system, such as valving the local water system to shut down mains for work and then reloading the mains before their return to use, can allow for contamination of the drinking water supply from backflow through unprotected domestic and fire connection services, which is covered under the section on cross connections and backflow.

Craun and Calderon (2001), in summarizing waterborne outbreaks from 1971 to 1998, found that of the 12 largest outbreaks caused by distribution system deficiencies, one in Indiana was associated with contamination of main interiors during storage in the pipe yard or on the street prior to pipe installation. This review also recalled the well-documented 1989 waterborne disease outbreak in Cabool, Missouri, which was associated with water meter repair and two large water main breaks during the winter (see Clark et al., 1991; Geldreich et al., 1992). The loss of system pressure and ineffective system controls allowed external contamination, such as sewage, to come in contact with the water meters and pipe interior. In addition, the insufficient use of best practices such as post-repair disinfection allowed the contamination to spread through the distribution system. Although yet to be verified, the EPA white paper, New and Repaired Water Mains (AWWA and EES, Inc., 2002e), states that about 5 percent of reported waterborne disease outbreaks in the United States over a 27-year period were associated with main construction and repair activities. Over 200,000 water main breaks occur every year (over 555 breaks per day) according to Kirmeyer et al. (1994). Over 4,000 miles of pipe are replaced every year and over 13,000 miles of new pipe are installed every year. These numbers, along with the fact that disease occurrence is underreported and contamination from such activities would be highly localized and undetectable in most cases, suggest that exposure to contaminated drinking water from main breaks and installation, repair, and replacement activities is likely to be significantly greater than has been documented.

ANSI/AWWA standards, particularly C600-99 for the installation of ductile iron mains and C651-99 for the disinfection of mains, are commonly employed to prevent microbial contamination during main rehabilitation and replacement (Pierson et al., 2001). However, the actual documentation and inspection of sanitary practices varies widely. Even well-run utility operations, for example, can experience a 30 percent failure rate in the approval of new mains based on water quality testing (Burlingame and Neukrug, 1993). Haas et al. (1998) reported that interior pipe surfaces are not free of microbial contaminants even under best case conditions. Thus, when a new main is installed or a valve is repaired, it is advisable to act as if some level of contamination has occurred to both the water and the materials. In both cases testing is required, and care should be taken to address potential contamination before the affected portion of the water system is returned to use. Prevention of contamination can also be facilitated by including the existing standards and additional training on sanitary practices in distribution system operator training requirements and sanitary survey guidelines. There is more variability in practices when it comes to preventing contamination during main breaks and failures than with the installation of new mains. Haas et al. (1998) reported that while 90 percent of utilities surveyed said that new mains must meet water quality criteria before they are released back to service, only 29 percent said samples were required to be collected in response to a main break. One common practice is to simply flush hydrants on the water mains in the affected area until the water runs clear. Because water main repairs are of varying complexity and occur under a variety of environmental conditions, and due to their unplanned nature may require quick response and return to service, the application of the same level of specifications used for new water mains may not be feasible.

The chemical and microbial contamination of distribution system materials and drinking water during mains breaks and during the installation, rehabilitation, and repair of water mains and appurtenances is a high priority issue. As discussed above, there have been many documented instances of significant health impacts from drinking water contamination associated with pipe failures and maintenance activities. The improved application of best practices, and operator training and certification, can reduce this risk.

FINISHED WATER STORAGE

Treated water storage facilities, of which there are 154,000 in the United States (AWWA, 2003), are of vital importance for drinking water distribution systems. Storage facilities are traditionally designed and operated to secure system hydraulic integrity and reliability, to provide reserve capacity for fire fighting and other emergencies, to equalize system pressure, and to balance water use throughout the day. To meet these goals, large volumes of reserve storage are usually incorporated into system operation and design, resulting in long detention times. Long detention times and improper mixing within such facilities provide an opportunity for both chemical and microbial contamination of the water. One of the most important manifestations of water quality degradation during water storage is a loss of disinfectant residual, which can be further compromised by temperature increases in water storage facilities under warm weather conditions. Internal chemical contamination can also occur due to leaching from coatings used in the storage facility, or solvents, adhesives, and other chemicals used to fabricate or repair floating covers. Until recently, water quality issues associated with such facilities have usually been considered as only secondary maintenance items such as cleaning and coating.

In addition to the internal degradation of water quality that occurs over time in water storage facilities, they are also susceptible to external contamination from birds, animals, wind, rain, and algae. This is most true for uncovered storage facilities, although storage facilities with floating covers are also susceptible to bacterial contamination due to rips in the cover from ice, vandalism, or normal operation. Even with covered storage facilities, contaminants can gain access through improperly sealed access openings and hatches or faulty screening of vents and overflows. The white paper, Finished Water Storage Facilities (AWWA and EES, Inc., 2002c), identified four waterborne disease outbreaks associated with covered storage tanks. In particular, in Gideon, Missouri, a *Salmonella typhimurium* outbreak occurred due to a bird contamination of a covered municipal water storage tank (Clark et al., 1996).

Water quantity and quality requirements in distribution storage management decisions are frequently in conflict. While water quantity objectives promote excessive storage, water quality objectives are geared toward minimizing residence times and frequent exercising of treated water facilities to maximize the stored water disinfectant residual. Appropriate balancing is therefore required to ensure disinfection effectiveness and sufficient level of service (Boulos et al., 1996; Hannoun and Boulos, 1997). Numerous standards prepared by ANSI/AWWA, Ten States Standards, and the National Sanitation Foundation (NSF) are available for the design, construction, and maintenance of water storage facilities. However, if retention times are long, disinfectant residual can drop, via reaction with oxidizable material in the water, to a level that is non-protective of microbial contamination. To minimize this potential problem, adequate turnover of the water in the facility is an essential operational parameter. It is also desirable to adequately mix (to eliminate dead zones) or prevent the short circuiting of the water entering and leaving the facility to shorten the water age in the facility.

The documented cases of waterborne disease outbreaks and the potential for contamination due to the large number of these facilities make this a high priority distribution system water quality maintenance and protection issue.

ADDITIONAL ISSUES OF CONCERN

Two distribution system issues not mentioned in the nine white papers that the committee believes are of significance to public health protection include the management of premise plumbing and the training of distribution system operators.

Premise Plumbing

Premise plumbing is that portion of the water distribution system from the main ferrule or water meter to the consumer's tap in homes, schools, hospitals, and other buildings. Virtually every problem identified in potable water transmission systems can also occur in premise plumbing. However, due to premise plumbing's higher surface area to volume ratio, longer stagnation times, and warmer temperatures (especially in the hot water system), the potential health threat can be magnified (Edwards et al., 2003). This is an important problem because it requires that individual homeowners be responsible for making decisions that will affect the safety of their drinking water. Premise plumbing is also a valuable asset, with more than 5.3 million miles of copper tube installed in buildings since 1963 (CDA, 2004). The estimated replacement value of premise plumbing in buildings is over 1 trillion dollars (Parsons et al., 2004) and the cost of a plumbing failure to an individual homeowner can exceeded $25,000. The problems of greatest concern within premise plumbing include microbial regrowth, leaching, permeation, infiltration, cross connections, leaks and the resulting indoor mold growth, scaling, and the high costs of failure.

Regrowth problems are exacerbated in premise plumbing due to very long stagnation times resulting in a loss of chlorine residual, to the presence of numerous microclimates, and to nutrient release from some pipes. Some clear links have been established between regrowth of opportunistic pathogens such as *Legionella* and *Mycobacterium* and waterborne disease amongst immunocompromised patients in hospitals (EPA, 2002b). A special concern is regrowth of pathogens within water heaters and showers, which may be exacerbated by recent efforts to reduce temperatures to prevent scalding and save energy (Spinks et al., 2003). The impact of low-flow shower fixtures on consumer exposure to airborne aerosols (a potential route for *Legionella* exposure—Blackburn et al., 2004) and endotoxins is unclear and in need of examination (see Rose et al., 1998, and Anderson et al., 2002).

A subset of the regrowth issue in homes deals with the presence of granular activated carbon in point-of-use treatment devices that can accumulate bacterial nutrients and neutralize disinfectant residuals, thus providing an ideal environment for microbial growth (Tobin et al., 1981; Geldreich et al., 1985; Reasoner et al., 1987; LeChevallier and McFeters, 1988). Several coliform bacteria (*Klebsiella, Enterobacter,* and *Citrobacter*) have been found to colonize granular activated carbon filters, regrow during warm-water periods, and discharge into the process effluent (Camper et al., 1986). The presence of a silver bacteriostatic agent did not prevent the colonization and growth of HPC bacteria in granular activated carbon filters (Tobin et al., 1981; Reasoner et al., 1987). Rogers et al. (1999) reported the growth of *Mycobacterium avium* in point-of-use filters in the presence of 1,000 µg/mL silver. While general microbial parameters such as HPC are also higher after point-of-use devices, there is currently no evidence that these microbes cause significant human health problems (WQA, 2003).

Leaching and permeation mechanisms within premise plumbing are the same as for public water supply transmission lines. However, the wide variety of materials used in building plumbing and associated treatment devices, the higher surface area to volume ratio, very long stagnation times, and lessened dilution increase the potential severity of the problem in premise plumbing. If permeation were to occur through the consumer's service line it would rarely be detected by routine monitoring of the distribution system. For new materials, levels of allowable contaminant leaching have been established through health based standards in ANSI/NSF Standard 61. However, ANSI/NSF Standard 61 is a voluntary test for premise plumbing, ANSI/NSF certification may not be required in all states for materials installed in buildings, and it is not clear whether states are applying materials tested under Standard 61 to premise plumbing. Indeed, there is reason to believe that certain existing NSF standards may not be sufficiently protective of public health in the context of lead leaching from in-line brass devices (Abhijeet, 2004).

Problems with infiltration and cross connections are also greater in premise plumbing than in the main distribution system. Premise plumbing is routinely subject to pressure and flow transients and is the site of minimum pressure in the distribution system. Moreover, the pressures measured in distribution systems during high flow events are an upper bound for pressures that occur in premise plumbing. Considering that direct or indirect cross connections are likely to be found in a majority of household plumbing systems (see previous discussion of backflow events and USC, 2002), and the known occurrence of leaks in premise service lines, there is significant potential for backflow and intrusion. Previous EPA white papers on intrusion and cross connections (LeChevallier et al., 2002; EPA, 2002b) do not explicitly include risks in premise plumbing. Thus, additional research is needed to determine the potential impacts of building hydraulics on intrusion and contamination from cross connections.

Other problems beyond the scope of the current effort, but which are nonetheless important, include reduced flow and high energy costs associated with scaling in pipes and water heaters, leak-induced mold growth in buildings, and the high cost of water damages and re-plumbing due to material failure.

The premise plumbing issue poses unique challenges because there is no obvious single party who could assume responsibility for the problem, which might be best addressed through changes in and enforcement of plumbing codes, third party standards, and public education. Utility involvement in overseeing solutions may be appropriate where distribution system water quality directly contributes to the problem, as is currently the case with the Lead and Copper Rule.

Distribution System Operator Training

Traditionally, training of drinking water distribution system operators has focused primarily on issues related to the mechanical aspects of water delivery (pumps and valves) and safety. However, system operators also are responsible for ensuring that the operation of the system from a quantity perspective does not cause degradation of water quality (EPA, 1999), and it is important that they receive adequate training to meet this need. The training should include an understanding of constituents that affect public health, such as disinfectants, disinfection by-products (DBPs), and metals, and how distribution system operations affect their concentrations. It should also include guidance on meeting water quality monitoring and reporting requirements, on how to interpret monitoring results, and on actions that should be taken when "positive" hits are detected (such as increased levels of coliforms or turbidity and decreased or depleted chlorine residuals). Most importantly, the distribution operator must be trained to make decisions regarding the proper balance of quality and quantity issues, such as in proper operation of distribution system storage facilities (Smith and Burlingame, 1994; Kirmeyer et al., 1999). The need to train operators is more pronounced in small systems where there are fewer staff members to aid operators in day-to-day decisions.

The need for the continuing and intensive training of operators of distribution systems has increased recently for three reasons. First, recent federal and state regulations (EPA, 1990) are more sophisticated and require enhanced skills for proper sample collection and preservation, as well as better understanding of aquatic chemistry and biology. Second, in many systems the new regulations (EPA, 1998) created a shift in the use of disinfectants in the distribution systems from a relatively simple application of chlorine to rather complicated application and maintenance of chloramines. Finally, with an increase in the importance of security of drinking water pipes, pumps, reservoirs, and hydrants, there is a corresponding increase in the responsibility of operators to make decisions during perceived security events.

4
Medium Priority Issues

Four of the white papers dealt with topics considered by the committee to have some potential impact on public health. These medium priority issues are biofilm growth, loss of disinfectant residual during nitrification and water aging, and intrusion. Control of these issues certainly should be considered as part of system-wide management.

BIOFILM GROWTH

Virtually every water distribution system is prone to the formation of biofilms, regardless of the purity of the water, type of pipe material, or the presence of a disinfectant. Growth of bacteria on surfaces can occur in the distribution system or in household plumbing. It is reasonably well documented that the suspended bacterial counts observed in distribution systems are the result of biofilm cell detachment rather than growth of organisms in the water (Characklis, 1988; Haudidier et al., 1988; van der Wende et al., 1989; van der Wende and Characklis, 1990). This phenomenon extends to autotrophic and heterotrophic organisms, coliforms (as noted in the TCR), and opportunistic pathogens. As a result of detachment, the biofilm can act as a continuous inoculum into finished water. The organisms can then be inhaled through bathing and showering or ingested.

The larger question is not whether biofilms are present, but whether biofilms are associated with disease. Biofilms in drinking water distribution systems are primarily composed of organisms typically found in the environment, and as such, are likely to be of limited health concern. These organisms are often enumerated by HPC, and yet epidemiological studies relating HPC counts to health effects are scarce due to their high cost and lack of funding, with only a few studies having been completed. In 1991, Calderon and Mood studied the impact of point-of-entry devices containing granular activated carbon on microbial water quality and health effects. Because granular activated carbon can enhance the growth of organisms detected by HPC, counts were elevated in the treated water. In this study, there was no correlation between elevated HPC and health effects. Other studies conducted by Payment et al. (1991a,b, 1997) in Canada suggested that there may be some association between distributed water, HPC counts, and gastroenteritis, but the findings were not clear.

Biofilms can be a reservoir for opportunistic pathogens. *M. avium* complex (*M. avium* and *M. intracellulare*; MAC) have been isolated from drinking water throughout the United States (Haas et al., 1983; duMoulin and Stottmeir, 1986; Carson et al., 1988; duMoulin et al., 1988; Fischeder et al., 1991; von Reyn et al., 1993, 1994; Glover et al., 1994). They have been implicated in infections and tuberculosis-like disease in the immunocompromised population, particularly those with AIDS (Horsburgh, 1991; Nightingale et al., 1992). The MAC is of such significant public health concern that the organisms are included on the EPA's Contaminant Candidate List (CCL). *Legionella* species have been shown to proliferate in drinking water (Wadowsky and Yee, 1983, 1985; Stout et al., 1985; Rogers et al., 1994). These bacteria can be found in water heaters, shower heads, and cooling towers, where their release can lead to respiratory disease in sensitive individuals. *Legionella* are specifically mentioned in the EPA's SWTR, with the maximum contaminant level goal set at zero. Both MAC and *Legionella* are more likely to proliferate in premise plumbing systems rather than the main distribution system.

The elimination of biofilms is essentially impossible and their control is difficult. Attempts to eliminate them are complicated by the observation that conditions that reduce the numbers of organisms of one type may potentially select for others. In general, biofilms can be managed by removing organic matter during water treatment that would support biofilm growth, judicious use of disinfectants, good distribution hygiene practices such as flushing, minimizing the corrosion of iron pipe surfaces, and managing contamination from external sources. Biofilm management is ideally accomplished by best practices that also reduce the magnitude of other water quality problems such as disinfection byproduct concentration, corrosion, and aesthetic concerns. This is true for both the utility owned distribution system and premise plumbing.

Biofilm growth is considered to be of medium priority because the potential for public health risk from this source of exposure is of lesser importance than phenomena included in the high risk category. The risks associated with biofilms appear to be most significant with opportunistic pathogens that may cause disease in the immunocompromised population. Because coliform regrowth in biofilms can lead to TCR violations, biofilm control can assist utilities in meeting the requirements of this regulation. Therefore, mechanisms for controlling biofilms may be of benefit both to reducing coliform levels and reducing the levels of opportunistic pathogens.

LOSS OF RESIDUAL VIA WATER AGE AND NITRIFICATION

This issue is related to an effect identified in two of the nine white papers—that of loss of disinfectant residual brought about by the aging of water and nitrification. Water age, which is considered to be synonymous with "hydraulic residence time," depends on both the physical characteristics of the system (such as flow rate, pipe size, configuration, and the amount of storage) as well as its mode of operation. From the point of entry into a distribution system to an individual consumer tap, water may be in transport for days to weeks. Systems that are "looped" may have shorter maximum water ages than systems that contain long pipelines with dead end sections. A distribution system is not generally uniform in structure but consists of a network of various elements having different physical, chemical, and biological characteristics such as differing size pipes and pipe materials, occurrence of pipe scales, and biofilms. Furthermore, some characteristics such as surface roughness may change with time, which in turn may influence the hydraulic residence time and the path water takes as it flows through the system.

Unlike specific degradative processes influencing water quality that are considered in this report, retention time is a characteristic that only indirectly affects water quality. Many degradative processes are time dependent and, therefore, more adversely affect water quality with increasing retention time. The degradative processes that are most influenced by residence time can be attributed to reactions occurring in the bulk water and at the pipe wall interface.

Biological nitrification is a process in which bacteria in distribution systems oxidize reduced nitrogen compounds (e.g., ammonia) to nitrite and then nitrate. It is an important process associated with nitrifying bacteria in distribution systems and long retention times in systems practicing chloramination. Like water age, it has a variety of direct and indirect effects on distribution system water quality.

In the committee's opinion, the most important problem exacerbated by both nitrification and by long retention times is loss of disinfectant residuals. Chlorine and chloramine loss during water aging is attributable to reactions with demand substances such as reduced iron in corrosion deposits, ammonia, and natural organic matter (NOM) both on the pipe surface and in the bulk phase. In so far as a residual in the distribution system is desirable, the microbial integrity of the system is compromised. Increased occurrence of microorganisms such as coliforms is associated with the loss of disinfectant residual (Wolfe et al., 1988, 1990). Similarly, the loss of chloramine residual driven by biological nitrification was deemed by the committee to be a significant health threat, and more important than the issue of high concentrations of nitrate and nitrite that result from nitrification. Indeed, a positive feedback loop between growth of nitrifying bacteria and chloramine loss can be established, since the loss of disinfectant residual removes one of the controls on the growth of these organisms.

The precise influence of water age on water quality is complex and clearly system specific, complicating potential control strategies. Water age, unlike other causes of distribution system water quality degradation, such as backflow, cannot be eliminated, only managed within the framework of numerous constraints. Water age is determined by flow rate and the internal volume of the distribution system network, and it can be estimated using tracer studies, mathematical models, system models, and computational fluid dynamics models. The physical aspects of pipe sizes and network layout are important considerations in minimizing water age. Research indicates that "dead ends" and low velocities should be avoided (AWWARF, 2004). This would favor the use of small diameter pipes and careful consideration of flow paths ("looped" geometry). Current design practice, however, typically dictates a design not based on water needed at the tap but on peak flows associated with fire fighting. This tends to result in a design incorporating comparatively large pipes with resulting lower flow rates. Network operation is also an important determinant of water age at any particular point in the system. Water may be routed to avoid excessively long residence times. Periodic flushing of system elements associated with long water age may also minimize water quality degradation by removal of pipe scales and sediment associated with disinfectant consumption and release of iron into the water. Finally, in the case of systems with multiple sources of supply, hydraulic modeling can be used to assess system operations to reduce maximum water age.

A strategy to control nitrification involves periodically switching to free chlorine, which is thought to reduce the active microbial population, or wholesale replacement of chloramines with free chlorine or chlorine dioxide. Other control strategies include reducing the ratio of chlorine to free ammonia, increasing pH, reducing the residence time by managing the flow, lowering the TOC levels in the distribution system via advanced treatment, and maintaining a fairly high residual as well as a low level of free ammonia.

The loss of disinfectant residual caused by increased water age and nitrification is considered a medium priority concern because it is an indirect health impact that compromises the biological integrity of the system and promotes microbial regrowth. See two sections below for further discussion of other, lower priority effects of water age and nitrification on distribution system water quality.

LOW PRESSURE TRANSIENTS AND INTRUSION

Ensuring safe distribution of treated water to consumers' taps requires, among other measures, protection from intrusion of contaminants into the distribution system during low pressure transients. Intrusion refers to the flow of non-potable water into drinking water mains through leaks, cracks, submerged air valves, faulty seals, and other openings resulting from low or negative pressures. Transient pressure regimes are inevitable; all systems will, at some time, be started up, switched off, or undergo rapid flow changes such as those caused by hydrant flushing, and they will likely experience the effects of human errors, equipment breakdowns, earthquakes, or other risky disturbances (Boulos et al., 2004; Wood et al., 2005). Transient events can have significant water quality and health implications. These events can generate high intensities of fluid shear and may cause resuspension of settled particles as well as biofilm detachment. Moreover, a low-pressure transient event, arising for example from a power failure or intermittent/interrupted supply, has the potential to cause the intrusion of contaminated groundwater into pipes with leaky joints or cracks. This is especially important in systems with pipes below the water table. Dissolved air (gas) can also be released from the water whenever the local pressure drops considerably, and this may promote the corrosion of steel and iron sections with subsequent rust formation and pipe damage. Even some common transient protection strategies, such as air relief valves or air chambers, if not properly designed and maintained may permit pathogens or other contaminants to find a "back door" route into the potable water distribution system. In the event of a large intrusion of pathogens, the chlorine residual normally sustained in drinking water distribution systems may be insufficient to disinfect contaminated water, which can lead to adverse health effects.

Low water pressure in distribution systems is a well-known risk factor for outbreaks of waterborne disease (Hunter, 1997), although there are limited field data (suggesting that additional field studies are needed). In 1997, a massive epidemic of multidrug-resistant typhoid fever (8,901 cases of typhoid fever and 95 deaths) was reported in the city of Dushanbe, Tajikistan, which affected about 1 percent of the city's population. Low water pressure and frequent water outages had contributed to widespread increases in contamination within the distribution system (Hermin et al., 1999). More recently (April 2002), a *Giardia* outbreak was reported at a trailer park in New York State causing six residents to become seriously ill (Blackburn et al., 2004). Contamination was attributed to a power outage, which created a negative pressure transient in the distribution system. This allowed contaminated water to enter the system through either a cross-connection inside a mobile home or through a leaking underground pipe that was near sewer crossings. During the same period (February 2001 to May 2002), a large-scale case-control study conducted in England of the risk factors for sporadic cryptosporidiosis suggested a strong association between self-reported diarrhea and reported low water pressure events (Hunter et al., 2005).

Intrusion events can be controlled or prevented by developing and implementing best distribution system operational practices such as the requirement for maintaining a sufficient water pressure and an adequate level of disinfectant residual throughout the distribution system, leak detection and control, replacement and rehabilitation of nearby sewer lines, proper hydrant and valve operations, redesign of air relief venting (above grade), routine monitoring program (a sudden drop in the chlorine residual could provide a critical indication to water system operators that there is a source of contamination in the system), use of transient modeling to predict and eliminate potential weak spots in the distribution system, and more rigorous applications of existing engineering standards.

Although there are currently insufficient data in the literature to indicate whether intrusion from pressure transients is a substantial source of risk to treated distribution system water quality, nevertheless intrusion is inherently a subset of backflow events, has health risks and is, therefore, an important distribution system water quality maintenance and protection issue.

5
Lower Priority Issues

Of the issues discussed in the white papers, four were felt to be of lower priority in terms of their potential for adverse health effects. Nonetheless, these issues, together with two additional issues, are important for maintaining a well-managed system, upholding high aesthetic quality, minimizing the energy required for distribution, and providing adequate quantities of water.

OTHER EFFECTS OF WATER AGE

As discussed above, water age has an indirect effect on water quality, with the most important being the reduction in disinfectant residual over time. However, there are a number of other alterations that may occur as water ages that merit discussion. First, with increasing age there can be increasing formation of DBPs (e.g., trihalomethanes and haloacetic acids). In-system production of some DBPs may be prevalent, for example, where pipe sediments contain significant organic matter and/or booster chlorination is practiced. There may also be increasing potential for nitrification with increasing water age, especially at higher temperatures. These latter effects of water age may be reduced by reducing the concentration of byproduct precursors (e.g., total organic carbon) and ammonia entering the distribution system.

The presence of high concentrations of corrosion products is frequently associated with long water age. Corrosion in distribution systems, as well as household plumbing, is a complex process still not adequately understood despite much research into the causes. A number of relevant water quality parameters such as disinfectant residual, redox potential, and pH are affected by water age, and these are believed to play an important role in the corrosion of pipe materials and the release of iron, copper, and lead from pipe scales, especially in low alkalinity waters. Control strategies are sometimes utilized such as changing the pH or adding phosphates to reduce lead and copper corrosion and the release of iron from pipe scales, but these measures are more effective if water age and the amount of stagnation are minimized. Other problems associated with water age include the development of objectionable taste and odors, water discoloration, and sediment accumulation.

Of these lower priority concerns, DBP formation and corrosion are the most important because of obvious health risks. Even so, the health risk of DBPs within a given system may be low compared to contaminants that have an acute health effect, and DBPs are covered by the Disinfectants/Disinfection By-Products Rule. While leached metals such as lead may be found at very high levels at the tap in some instances, the relationship of their concentrations to water age is not yet adequately understood.

OTHER EFFECTS OF NITRIFICATION

As discussed above, nitrification is a process carried out by ammonia-oxidizing bacteria in the environment that produces nitrite and nitrate, and thus occurs whenever the substrate (ammonium) is present in the waters. There exist abundant data on the impact of nitrate and nitrite on public health, especially on methemoglobinemia (blue baby syndrome, an acute response to nitrite that results in a blockage of oxygen transport—Bouchard et al., 1992). It affects primarily infants below six months of age, but it may occur in adults of certain ethnic groups (Navajos, Eskimos) and those suffering from a genetic deficiency of certain enzymes (Bitton, 1994). Pregnant women may also be at a higher risk of methemoglobinemia than the general population (Bouchard et al., 1992).

Nitrate levels may be important under certain conditions, although the relative source contribution from drinking water is expected to be a maximum of about 1–2 mg/L as nitrogen and typically would be much less that this. Numerous papers have focused on the impact of nitrate nitrogen (nitrate plus nitrite) in drinking waters (Sandor et al., 2001; Gulis et al., 2002; Kumar et al., 2002; De Roos et al., 2003; Coss et al., 2004; Fewtrell, 2004). However, the concentration at which nitrate nitrogen in drinking waters presents a health risk is unclear (Fewtrell, 2004). Nitrate may be reduced to nitrite in the low pH environment of the stomach, reacting with amines and amides to form N-nitroso compounds (Bouchard et al., 1992; De Roos et al., 2003). Nitrosamines and nitrosamides have been linked to different types of cancer, but the intake of nitrate from drinking water and its causal relation to the risk of cancer is still a matter of debate (Bouchard et al., 1992). A study by Gulis et al. (2002) in Slovakia related increased colorectal cancer and non-Hodgkins lymphoma to medium (10.1–20 mg/l) and high (20.1–50 mg/l) concentrations of nitrate nitrogen in drinking waters. Similarly, Sandor et al. (2001) showed a correlation between the consumption of waters containing greater than 88 mg/l nitrate nitrogen and gastric cancer.

Current nitrite and nitrate MCLs, which are regulated at the entry point to the distribution system, have been set at 1 and 10 mg/l as nitrogen, respectively, in the United States and Canada. The World Health Organization recommends 11.3 mg/l nitrate nitrogen as a guideline value. van der Leeden et al. (1990) presented data up to 1962 in which 93 percent of all U.S. water supplies contain less than 5 mg/l nitrate (it was not specified if the concentrations were nitrate nitrogen or nitrate). However, this may be changing as a result of the increased use of nitrate-containing fertilizers. It has also been shown that chloramination, which is on the increase as an alternative disinfectant, may result in elevated levels of nitrate in waters because of partial nitrification (Bryant et al., 1992), but the increment in nitrate plus nitrite nitrogen from this source would typically be less than 1 mg/L. Information obtained from the ICR database indicates that up to 65 percent of the surface water systems in the United States may use chloramination in the near future (up from 33 percent currently) (EPA M/DBP FACA Support Document, 2000). This may have the unintended result of a possible increase in the final concentration of nitrate in drinking water. In most cases, the current MCL seems to be well below the concentrations at which risk has been observed. However, some special populations (pregnant women, infants) as well as some ethnic groups may more susceptible to adverse health effects as a result of elevated nitrate concentrations in drinking waters (Bitton, 1994; De Roos et al., 2003).

Lesser effects are that nitrification in low alkalinity waters can reduce alkalinity and decrease the pH. This may cause the pH to decrease to the point that corrosion of lead or copper becomes a problem.

The formation of nitrate and nitrite is considered a relatively low priority concern for distribution systems compared to the other concerns mentioned in this report, primarily because the amount of nitrate generated would likely be less than 10 percent of the MCL. Furthermore, except in very special situations drinking water is not a major source of these substances in the average diet. For example, nitrate is especially abundant in many leafy green vegetables.

PERMEATION

Permeation in water distribution systems occurs when contaminants external to the pipe materials and non-metallic joints pass through these materials into the drinking water. Permeation is generally associated with plastic non-metallic pipes (Holsen et al., 1991). The contaminants that are most commonly found to permeate plastic pipes are organic chemicals that are lipophilic and non-polar such as highly volatile hydrocarbons and organic solvents (Holsen et al., 1991; Burlingame and Anselme, 1995). These chemicals can readily diffuse through the plastic pipe matrix, alter the plastic material, and migrate into the water within the pipe.

The most common example of permeation of water mains and fittings is associated with soil contamination of the area within which the pipe was placed (Glaza and Park, 1992). The majority of permeation incidents appear to be associated with gasoline related organic chemicals. These incidents have occurred at high-risk sites, such as industrial sites and near underground chemical storage tanks, as well as at lower risk residential sites (Holsen et al., 1991). In some cases the integrity of the pipe has been irreversibly compromised, requiring the complete replacement of the contaminated section.

Although there is the potential for water quality degradation as a result of the permeation of plastic pipe, especially in the water's taste and odor aspects, the health impacts associated with such permeation are not expected to be significant. In some permeation incidents, the concentrations of certain chemicals have been shown to reach levels in the low parts per million, which are well above their respective MCLs (AWWA and EES, Inc., 2002a.). However, these MCLs are based on long-term exposure, and the short-term risk levels for these chemicals are generally much higher. In the case of permeation by gasoline components, the taste or odor thresholds of the majority of these chemicals are below the levels that would pose a short-term risk (EPA, 2002e,f,g,h). In addition, these high concentrations would be expected to occur during worst case situations where water has been in contact with the affected pipe for a considerable length of time. During periods of normal water use these concentrations would expected to be much lower.

Appropriate measures can be taken to minimize the occurrence of permeation, such as issuing regulations or guidelines that define the conditions under which plastic pipe should be used. For example, the State of California precludes the use of plastic pipe in areas subject to contamination by petroleum distillates (California Code of Regulations, Title 22, Division 4, Chapter 16, Article 5, Section 64624f).

After assessing the potential health impacts associated with permeation, the committee has concluded that the potential health impacts are low and that distribution systems can best be protected through measures that minimize the conditions under which permeation can occur.

LEACHING

All materials in the water distribution system undergo reactions that introduce substances to the water via a process known as "leaching." Pipes, fittings, linings, and other materials used in joining or sealing pipes leach at least some substances to water through corrosion, dissolution, diffusion, or detachment. Internal coatings in water storage facilities can also leach substances. Most substances leaching to water from materials in the distribution system do not pose a public health threat due the fact they are nontoxic, present only at trace levels, or are in a form unlikely to cause health problems. Taste and odor complaints are possible, however (see Choi et al., 1994, and Khiari et al., 2002, for examples).

Under some circumstances, leaching of toxic contaminants occurs at levels that pose a substantial health threat. PVC pipes manufactured before about 1977 are known to leach carcinogenic vinyl chloride into water at levels above the MCL (AWWA and EES, Inc., 2002a). It should be noted that the MCL is based on a measurement of samples at the treatment plant and not within the distribution system. To protect against a health problem from this source, sampling in the distribution system would have to be required after installation of new PVC pipe. Cement materials have, under unusual circumstances, leached aluminum to drinking water at concentrations that caused death in hemodialysis and other susceptible patients (Berend et al., 2001). Because levels of aluminum normally present in drinking water can also threaten this population, the FDA has issued guidance for water purification pre-treatments in the U.S. for dialysis and other patients (Available on-line at *http://www.gewater.com/library/tp/ 1111_Water_The.jsp*). Finally, excessive leaching of organic substances from linings, joining and sealing materials have occasionally been noted in the literature, and asbestos fibers in water are regulated with an MCL. Potential problems with lead and copper leaching to water are managed through the EPA via the LCR and, thus, are not considered further here.

Problems from older distribution system materials can be managed by monitoring of contaminant leaching in the distribution system, adjustments to water chemistry, or by costly replacement of the material. Lead leaching to water from old lead pipe is managed in this way via the LCR. For new materials, the NSF establishes levels of allowable contaminant leaching through ANSI/NSF Standard 61. It should be noted that ANSI/NSF Standard 61, which establishes minimum health effect requirements for chemical contaminants and impurities, does not establish performance, taste and odor, or microbial growth support requirements for distribution system components. Research has shown that distribution system components can significantly impact the microbial quality of drinking water via leaching. For example, pipe gaskets and elastic sealants (containing polyamide and silicone) can be a source of nutrients for bacterial proliferation. Colbourne et al. (1984) reported that *Legionella* were associated with certain rubber gaskets. Organisms associated with joint-packing materials include populations of *Pseudomonas aeruginosa, Chromobacter spp., Enterobacter aerogenes,* and *Klebsiella pneumoniae* (Schoenen, 1986; Geldreich and LeChevallier, 1999). Coating compounds for storage reservoirs and standpipes can contribute organic polymers and solvents that may support regrowth of heterotrophic bacteria (Schoenen, 1986; Thofern et al., 1987). Liner materials may contain bitumen, chlorinated rubber, epoxy resin, or tar-epoxy resin combinations that can support bacterial regrowth (Schoenen, 1986). PVC pipes and coating materials may leach stabilizers that can result in bacterial growth. Studies performed in the United Kingdom reported that coliform isolations were four times higher when samples were collected from plastic taps than from metallic faucets (cited in Geldreich and LeChevallier, 1999). Although procedures are available to evaluate growth stimulation potential of different materials (Bellen et al., 1993), these tests are not applied in the United States by ANSI/NSF. Standards or third-party certification that establishes performance, taste and odor, or microbial growth support requirements for distribution system components could be considered. In spite of these limitations and occasional problems, it is currently believed that leaching is a relatively low priority relative to other distribution system problems.

ADDITIONAL ISSUES OF CONCERN

Control of Post Precipitation

Control of post precipitation in distribution systems is an important part of any program to control the quality of water in distribution systems. Post precipitation can result from introduction of water to distribution systems that is super-saturated with calcium carbonate, from introduction of a phosphate corrosion inhibitor into the filter effluent of an alum coagulation plant creating an aluminum phosphate precipitate, from water that is supersaturated with aluminum hydroxide, from water that is supersaturated with selected silicate minerals, as well as other causes. Post-precipitation causes an increase in pipe roughness and a decrease in effective pipe diameter, resulting in loss of hydraulic capacity accompanied by an increase in energy required to distribute water, in production of biofilms, and in deterioration of the aesthetic quality of tap water. If the material is loosely attached to the pipe wall, such as some aluminum precipitates, hydraulic surges can result in substantial increases in the turbidity of tap water. Treatment of water to avoid excessive post-precipitation thus is an important asset management issue. It is not amenable to regulation, but it is an important part of the guidance that should accompany distribution system regulations.

6
Summary

The purpose of this report was to review published material in order to identify trends relevant to the deterioration of drinking water quality in water supply distribution systems and to identify and prioritize issues of greatest concern. The trends relevant to the deterioration of drinking water quality in distribution systems include:

- **Aging distribution systems.** Increasing numbers of main breaks and pipe replacement activities are a possibility as systems age, depending on the pipe materials and linings used, the water quality, and system operation and maintenance practices.
- **Decreasing numbers of waterborne outbreaks reported per year since 1982, but an increasing percentage attributable to distribution system issues.** This trend is probably is related to better treatment of surface water.
- **Increasing host susceptibility to infection and disease in the U.S. population.** This trend is caused by aging of the U. S. population, the increase in the incidence of AIDS, and the increasing use of immunosuppressive therapy.
- **Increasing use of bottled water and point of use treatment devices.** This trend suggests that exposure to tap water on a per capita basis may be decreasing. However, it should be kept in mind that point-of-use devices can support microbial regrowth.

The issues from the nine EPA white papers have been prioritized using categories of highest, medium, and lower priority. Also, several significant issues that were overlooked in previous reports have been identified by the committee and added. The highest priority issues are those that have a recognized health risk based on clear epidemiological and surveillance data.

Cross connections and backflow. Cross connections and backflow events are ranked as the highest priority because of the long history of recognized health risks posed by cross connections, the clear epidemiological and surveillance data implicating these events with outbreaks or sporadic cases of waterborne disease, and the availability of proven technologies to prevent cross connections.

Contamination during installation, rehabilitation, and repair of water mains and appurtenances. Chemical and microbial contamination of distribution system materials and drinking water during mains breaks and during the installation, rehabilitation, and repair of water mains and appurtenances is a high priority issue because there have been many documented instances of significant health impacts from drinking water contamination associated with pipe failures and maintenance activities.

Improperly maintained and operated distribution system storage facilities. Several documented waterborne disease outbreaks and the potential for contamination due to the large number of these facilities makes this a high priority distribution system water quality maintenance and protection issue.

Control of water quality in premise plumbing. Virtually every problem identified in potable water transmission systems can also occur in premise plumbing, and some are magnified because of premise plumbing characteristics and the way in which water is used in residences. Health risks associated with premise plumbing are hard to assess because the majority of health problems are likely to be sporadic, unreported cases of waterborne disease that affect single households. Waterborne disease outbreaks due to premise plumbing failures in residential buildings have been documented.

Distribution system operator training. Training of drinking water distribution system operators traditionally has focused on issues related to the mechanical aspects of water delivery (pumps and valves) and safety. System operators are also responsible for ensuring that conveyance of the water does not allow degradation of water quality, and it is important that they receive adequate training to meet this responsibility.

Medium priority issues are those where existing data suggest that the health risks are low or limited to sensitive populations. Issues where there were insufficient data to determine the magnitude of the health risk were also classified as medium priority.

Biofilm Growth. Although biofilms are widespread in distribution systems, the public health risk from this source of exposure appears to be limited to opportunistic pathogens that may cause disease in the immunocompromised population. Some data suggest that biofilms may protect microbial pathogens from disinfection, but there are few studies directly linking health effects to biofilms.

Loss of Disinfectant Residual. The loss of disinfectant residual caused by increased water age and nitrification is considered a medium priority concern because it is an indirect health impact that compromises the biological integrity of the system and promotes microbial regrowth.

Intrusion. Intrusion from pressure transients is a subset of the cross-connection and backflow issue. It has associated health risks, and is therefore an important distribution system water quality maintenance and protection issue. There are insufficient data to indicate whether it is a substantial health risk, however.

Lower priority issues are those that are already covered by current regulations, well-managed in the majority of water distribution systems, and are unlikely to pose a health risk.

Other Effects of Water Age. The quality of distributed water, in particular water age, also has indirect effects such as (1) DBP formation in distribution systems with increasing water age that might cause the MCLs for these substances to be exceeded and (2) enhanced corrosion and the release of metals from corrosion scales. DBPs and common corrosion products are covered by the D/DBPR and the LCR, respectively.

Nitrification. Nitrification that results in (1) the formation of nitrite and nitrate in quantities that cause the MCLs for these substances to be exceeded or (2) the release of excessive concentrations of metal ions should be avoided. However, the formation of nitrate and nitrite is considered a relatively low risk for distribution systems compared to the other concerns mentioned in this report.

Permeation. Permeation of chemicals through plastic pipe can occur, but the potential health impacts are low and distribution systems can best be protected through measures that minimize the conditions under which permeation can occur.

Leaching. Excessive leaching of organic substances from pipe materials, linings, joining and sealing materials, coatings, and cement mortar pipe have occasionally been noted in the literature. Leaching is a relatively low priority relative to other distribution system problems and can be controlled by regulating the materials that are used in distribution and premise plumbing systems, by specifying the water chemistry that must be used if certain materials are to be employed, and by appropriate monitoring requirements.

Post-precipitation. An additional issue of lower priority is the control of post-precipitation, which causes an increase in pipe roughness and a decrease in effective pipe diameter, resulting in loss of hydraulic capacity accompanied by an increase in the energy required to distribute water, in the production of biofilms, and in the deterioration of tap water's aesthetic quality.

Deteriorating infrastructure was not included as one of the issues that the committee prioritized because it is the ultimate cause of many of the other events that are discussed in this report, such as:

- water main breaks and contamination that results during their repair,
- contamination from decaying storage structures and their inadequate maintenance,
- intrusion and water loss,
- occurrence of excessive biofilms and nitrification,
- system design and operation practices that cause the water quality to degrade, and
- excessive deposits from corrosion and post-precipitation.

Solutions to problems caused by deteriorating infrastructure are thus expected to be applicable to most of the problems already discussed in this report. It should be noted that the rate of degradation of distribution system materials will vary from system to system depending on water quality and system operation and maintenance practices, such that the relationship between the age of a given system, its state of deterioration, and risk cannot be easily predicted. Confronting deteriorating infrastructure requires good asset management, including procedures to monitor and assess the condition of the distribution system and water quality changes that occur during distribution. Furthermore, appropriate maintenance, repair, and replacement should be carried out as needed, and operating and capital budgets should be available to finance this work.

References

Abhijeet, D. 2004. Reconsidering lead corrosion in drinking water: product testing, direct chloramine attack and galvanic corrosion. Virginia Tech MS Thesis.

American Water Works Association (AWWA). 2001. Reinvesting in drinking water structure: dawn of the replacement era. Denver, CO: AWWA.

AWWA. 2003. Water Stats 2002 Distribution Survey CD-ROM. Denver, CO: AWWA.

AWWA and EES, Inc. 2002a. Permeation and leaching. Available on-line at *http://www.epa.gov/safewater/tcr/pdf/permleach.pdf.* Accessed March 16, 2005.

AWWA and EES, Inc. 2002b. Effects of water age on distribution system water quality. *http://www.epa.gov/safewater/tcr/pdf/waterage.pdf.* Accessed March 16, 2005.

AWWA and EES, Inc. 2002c. Finished water storage facilities. Available on-line at *http://www.epa.gov/safewater/tcr/pdf/storage.pdf.* Accessed March 16, 2005.

AWWA and EES, Inc. 2002d. Nitrification. Available on-line at *http://www.epa.gov/safewater/tcr/pdf/nitrification.pdf.* Accessed March 16, 2005.

AWWA and EES, Inc. 2002e. New or repaired water mains. Available on-line at *http://www.epa.gov/safewater/tcr/pdf/maincontam.pdf.* Accessed March 16, 2005.

American Water Works Association Research Foundation (AWWARF). 2004. Managing distribution retention time to improve water quality: phase I. Report no. 91006F (RFP 2769). Denver, CO: Binnie Black and Veatch and AWWARF.

American Water Works Service Co., Inc. (AWWSC). 2002. Deteriorating buried infrastructure management challenges and strategies. Available on-line at *http://www.epa.gov/safewater/tcr/pdf/infrastructure.pdf.* Accessed March 16, 2005.

Anadu, E. C., and A. K. Harding. 2000. Risk perception and bottled water use. J. Amer. Water Works Assoc. 92 (11):82–92.

Anderson, W. B., R. M. Slawson, and C. I. Mayfield. 2002. A review of drinking-water-associated endotoxin, including potential routes of human exposure. Can. J. Microbiol. 48:567–587.

Bellen, G. E., S. H. Abrishami, P. M. Colucci, and C. Tremel. 1993. Methods for assessing the biological growth support potential of water contact materials. Denver, CO: AWWARF.

Berend, K., G. Van Der Voet, and W. H. Boer. 2001. Acute aluminum encephalopathy in a dialysis center caused by a cement mortar water distribution pipe. Kidney International 59(2):746–753.

Bitton, G. 1994. Role of microorganisms in biogeochemical cycles. Pp. 51–73 In Wastewater Microbiology. New York: John Wiley & Sons, Inc.

Blackburn, B. G., G. F. Craun, J. S. Yoder, V. Hill, R. L. Calderon, N. Chen, S. H. Lee, D. A. Levy, and M. J. Beach. 2004. Surveillance for waterborne-disease outbreaks associated with drinking water—United States, 2001–2002. MMWR 53(SS-8):23-45.

Bouchard, D. C., M. K. Williams, and R. Y. Surampalli. 1992. Nitrate contamination of groundwater: sources and potential health effects. J. Amer. Water Works Assoc. 84(9):84–90.

Boulos, P. F., W. M. Grayman, R. W. Bowcock, J. W. Clapp, L. A. Rossman, R. M. Clark, R. A. Deininger, and A. K. Dhingra. 1996. Hydraulic mixing and free chlorine residual in reservoirs. J. Amer. Water Works Assoc. 88(7):48–59.

Boulos, P. F., K. E. Lansey, and B. W. Karney. 2004. Comprehensive Water Distribution Systems Analysis Handbook for Engineers and Planners. Pasadena, CA: MWH Soft Pub.

Bryant, E. A., G. P. Fulton, and G. C. Budd. 1992. Chloramination. Pp. 128-170 In Disinfection Alternatives for Safe Drinking Water. New York: Van Nostrand Reinhold.

Burlingame, G. A., and H. M. Neukrug. 1993. Developing proper sanitation requirements and procedures for water main disinfection. In Proceedings of AWWA Annual Conference. Denver, CO: AWWA.

Burlingame, G. A., and C. Anselme. 1995. Distribution system tastes and odors. In Advances in Taste-and-Odor Treatment and Control. AWWA Research Foundation Cooperative Research Report. Denver, CO: AWWARF.

Calderon, R. L., and E. W. Mood. 1991. Bacteria colonizing point-of-entry, granular activated carbon filters and their relationship to human health. CR-813978-01-0. Washington, D.C.: EPA.

Camper, A. K., M. W. LeChevallier, S. C. Broadway, and G. A. McFeters. 1986. Bacteria associated with granular activated carbon particles in drinking water. Appl. Environ. Microbiol. 52:434–438.

Carson, L. A., L. A. Bland, L. B. Cusick, M. S. Favero, G. A. Bolan, A. L. Reingold, and R. A. Good. 1988. Prevalence of nontuberculous mycobacteria in water supplies of hemodialysis centers. Appl. Environ. Microbiol. 54:3122–3125.

CDA. 2004. Copper facts. Available on-line at *http://www.copper.org/education/c-facts/c-plumbing.html*. Accessed February 11, 2005.

Centers for Disease Control and Prevention. 2004. Cancer survivorship—United States, 1971–2001. MMWR 53(24):526-529.

Characklis, W. G. 1988. Bacterial regrowth in distribution systems. Denver, CO: AWWARF.

Choi, J., M. Fadel, L. Gammie, J. Rahman, and J. Paran. 1994. Sniff new mains...before customers complain. Opflow 20:10:3.

Clark, R. M., W. M. Grayman, J. A. Good rich, R. A. Deininger, and A. F. Hess. 1991. Field testing distribution water quality models. J. Amer. Water Works Assoc. 83(7):67–75.

Clark, R. M., E. E. Geldreich, K. R. Fox, E. W. Rice, C. H.Johnson, J. A. Goodrich, J. A. Barnick, and F. Abdesaken. 1996. Tracking a *Salmonella* serovar *typhimurium* outbreak in Gideon, Missouri: role of contaminant propagation modelling. Journal of Water Supply Research and Technology-Aqua. 45:171–183.

Colbourne, J. S., D. J. Pratt, M. G. Smith, S. P. Fisher-Hoch, and D. Harper. 1984. Water fittings as sources of *Legionella pneumophila* in hospital plumbing system. Lancet i:210–213.

Craun, G. F., and R. L. Calderon. 2001. Waterborne disease outbreaks caused by distribution system deficiencies. J. Amer. Water Works Assoc. 93(9):64–75.

Coss, A., K. P. Cantor, J. S. Reif, C. F. Lynch, and M. H. Ward. 2004. Pancreatic cancer and drinking water and dietary sources of nitrate and nitrite. Amer. J. Epidemiol. 159(7):693–701.

De Roos, A. J., M. H. Ward, C. F. Lynch, and K. P. Cantor. 2003. Nitrate in public water supplies and the risk of colon and rectum cancers. Epidemiology 14(6):640–649.

duMoulin, G. C., K. D. Stottmeier, P. A. Pelletier, T. A. Tsang, and J. Hedley-Whyte. 1988. Concentration of *Mycobacterium avium* by hospital hot water systems. J. Amer. Medical Assoc. 260:1599–1601.

duMoulin, G. C., and K. D. Stottmeir. 1986. Waterborne mycobacteria: an increasing threat to health. American Society for Microbiology News 52:525–529.

Edwards, M., D. Bosch, G. V. Loganathan, and A. M. Dietrich. 2003. The Future Challenge of Controlling Distribution System Water Quality and Protecting Plumbing Infrastructure: Focusing on Consumers. Presented at the IWA Leading Edge Conference in Noordwijk, Netherlands. May 2003.

Environmental Protection Agency (EPA). 1990. Fact sheet: drinking water regulations under the Safe Drinking Water Act. Washington, DC: EPA Office of Drinking Water Criteria and Standards Division.

EPA. 1998. National Primary Drinking Water Regulations: disinfectants and disinfection by-products; Final Rule. Federal Register 63(241).

EPA. 1999. Guidelines for the certification and recertification of the operators of community and nontransient noncommunity public water systems. Federal Register 64(24).

EPA. 2002a. 2000 Community water system survey. EPA 815-R-02-005A. Washington, D.C.: EPA Office of Water.

EPA. 2002b. Potential contamination due to cross-connections and backflow and the associated health risks: an issues paper. Available on-line at *http://www.epa.gov/safewater/tcr/pdf/ccrwhite.pdf.* Accessed March 16, 2005.

EPA.. 2002c. The clean water and drinking water infrastructure gap analysis. Washington, D.C.: EPA.

EPA. 2002d. Health risks from microbial growth and biofilms in drinking water distribution systems. Available on-line at *http://www.epa.gov/safewater/tcr/pdf/biofilms.pdf.* Accessed on March 16, 2005. Washington, DC: EPA.

EPA. 2002e. Technical fact sheet on: Benzene. Available on-line at *http://www.epa.gov/OGWDW/dwh/t-voc/benzene.html.* Accessed on March 16, 2005.

EPA. 2002f. Technical fact sheet on: xylenes. Available on-line at *http://www.epa.gov/OGWDW/dwh/t-voc/xylenes.html.* Accessed on March 16, 2005.

EPA. 2002g. Technical fact sheet on: Toluene. Available on-line at *http://www.epa.gov/OGWDW/dwh/t-voc/toluene.html.* Accessed on March 16, 2005.

EPA. 2002h. Technical fact sheet on: Ethylbenzene. Available on-line at *http://www.epa.gov/OGWDW/dwh/t-voc/ethylben.html.* Accessed on March 16, 2005.

EPA. 2004. Factoids: Drinking water and ground water statistics for 2003. EPA 816-K-03-001. Washington, D.C.: EPA Office of Water.

Fewtrell, L. 2004. Drinking-water nitrate, methemoglobinemia, and global burden of disease: as discussion. Environmental Health Perspectives 112(14):1371–1374.

Fischeder, R. R., R. Schulze-Robbecke, and A. Weber. 1991. Occurrence of mycobacteria in drinking water samples. Zbl. Hygiene 192:154–158.

Frost, F. J., G. F. Craun, and R. L. Calderon. 1996. Waterborne disease surveillance. J. Amer. Water Works Assoc. 88(9):66-75.

Geldreich, E. E., R. H. Taylor, J. C. Blannon, and D. J. Reasoner. 1985. Bacterial colonization of point-of-use water treatment devices. J. Amer. Water Works Assoc. 77(2):72–80.

Geldreich, E. E., K. R. Fox, J. A. Goodrich, E. W. Rice, R. M. Clark, D. L. and Swerdlow. 1992. Searching for a water supply connection in the Cabool, Missouri disease outbreak of *Escherichia coli* 0157:H7. Water Research 26(8):1127–1137.

Geldreich, E. E., and M. W. LeChevallier. 1999. Microbial water quality in distribution systems. Pp. 18.1–18.49 In: Water Quality and Treatment, 5th edition. R. D. Letterman (ed.). New York: McGraw-Hill.

Gerba, C. P., J. B. Rose, et al. 1996. Sensitive populations: Who is at the greatest risk? International Journal of Food Microbiology 30(1-2):113-123.

Glaza, E. C., and J. K. Park. 1992. Permeation of organic contaminants through gasketed pipe joints. J. Amer. Water Works Assoc. 84(7):92–100.

Glover, N. A., N. Holtzman, T. Aronson, S. Froman, O. G. W. Berlin, P. Dominguez, K. A. Kunkel, G. Overturf, G. Stelma, Jr., C. Smith, and M. Yakrus. 1994. The isolation and identification of *Mycobacterium avium* complex (MAC) recovered from Los Angeles potable water, a possible source of infection in AIDS patients. International J. Environ. Health Res. 4:63–72.

Guilaran, Y.-T. 2004. EPA Presentation to the NRC Committee of Public Water Supply Distribution Systems on October 27, 2004.

Gulis, G., M. Czompolyova, and J. R. Cerhan. 2002. An ecologic study of nitrate in municipal drinking water and cancer incidence in Trnava District, Slovakia. Environmental Research 88(3):182–187.

Haas, C. N., M. A. Meyer, and M. E. Paller. 1983. The ecology of acid-fast organisms in water supply, treatment, and distribution systems. J. Amer. Water Works Assoc. 75:39–144.

Haas, C. N., R. B. Chitluru, M. Gupta, W. O. Pipes, and G. A. Burlingame. 1998. Development of disinfection guidelines for the installation and replacement of water mains. Denver, CO: AWWARF.

Hannoun, I. A., and P. F. Boulos. 1997. Optimizing distribution storage water quality: a hydrodynamic approach. Journal of Applied Mathematical Modeling 21(8):495–502.

Hasit, Y. J., A. J. DeNadai, H. M. Gorrill, R. S. Raucher, and J. Witcomb. 2004. Cost and benefit analysis of flushing. Denver, CO: AWWA Research Foundation.

Haudidier, K, J. L. Paquin, T. Francais, P. Hartemann, G. Grapin, F. Colin, M. J. Jourdain, J. C. Block, J. Cheron, O. Pascal, Y. Levi, and J. Miazga. 1988. Biofilm growth in drinking water network: a preliminary industrial pilot plant experiment. Water Sci. Technol. 20:109.

Hermin, J. H. R. Villar, J. Carpenter, L. Roberts, A. Samaridden, L. Gasanova, S. Lomakina, C. Bopp, L. Hutwagner, P. Mead, B. Ross, and E. D. Mintz. 1999. A massive epidemic of multidrug-resistant typhoid fever in Tajikistan associated with consumption of municipal water. Journal of Infectious Diseases 179:1416–1422.

Holsen, T. M., J. K. Park, D. Jenkins, and R. E. Selleck. 1991. Contamination of potable water by permeation of plastic pipe. J. Amer. Water Works Assoc. 83(8):53–56.

Horsburgh, C. R. 1991. *Mycobacterium avium* complex infection in the acquired immunodeficiency syndrome. New England Journal of Medicine 324:1332–1338.

Hunter, P. R. 1997. Waterborne disease: epidemiology an ecology. Chichester, UK: Wiley.

Hunter, P. R., R. M. Chalmers, S. Hughes, and Q. Syed. 2005. Self-reported diarrhea in a control group: a strong association with reporting of low-pressure events in tap water. Clinical Infectious Diseases 40:e32–34.

ICF Consulting, Inc. 2004. Exposure assessment of pathogens and toxic chemicals in drinking water distribution systems workshop. Washington, D.C.: EPA.

Khiari, D., S. Barrett, R. Chinn, A. Bruchet, P. Piriou, L. Matia, F. Ventura, I. Suffet, T. Gittelman, and P. Luitweiler. 2002. Distribution generated taste-and-odor phenomena. Denver, CO: AWWARF.

Kirmeyer, G., W. Richards, and C. D. Smith. 1994. An assessment of water distribution systems and associated research needs. Denver, CO: AWWARF.

References

Kirmeyer, G. J., L. Kirby, B. M. Murphy, P. F. Noran, K. D. Martel, T. W. Lund, J. L. Anderson, and R. Medhurst. 1999. Maintaining and operating finished water storage facilities. Denver, CO: AWWARF.

Kirmeyer, G. K., M. Freidman, K. Martel, D. Howie, M. LeChevallier, M. Abbaszadegan, M. Karim, J. Funk, and J. Harbour. 2001. Pathogen intrusion into the distribution system. Denver, CO: AWWARF.

Kumar, S., A. B. Gupta, and S. Gupta. 2002. Need for revision of nitrates standards for drinking water: a case study of Rajasthan. Indian Journal of Environmental Health 44(2):168–172.

LeChevallier, M. W., and G. A. McFeters. 1988. Microbiology of activated carbon. Pp. 104-119 In: Drinking Water Microbiology, Progress and Recent Developments. G. A. McFeters (ed.). New York: Springer-Verlag.

LeChevallier, M., R. Gullick, and M. Karim. 2002. The potential for health risks from intrusion of contaminants into the distribution system from pressure transients. Draft Distribution System White Paper. Washington, D.C.: EPA

Lee, S. H., D. A. Levy, G. F. Craun, M. J. Beach, and R. L. Calderon. 2002. Surveillance for waterborne-disease outbreaks in the United States, 1999–2000. MMWR 51(No. SS-8):1–49.

Mackey, E. D., J. Davis, L. Boulos, J. C. Brown, and G. F. Crozes. 2003. Consumer perceptions of tap water, bottled water, and filtration devices. Denver, CO: AWWARF.

Morris J. G., and M. Potter. 1997. Emergence of new pathogens as a function of changes in host susceptibility. Emerg. Inf. Dis. 3(4):435–441

National Research Council (NRC). 1999. Watershed Management for Potable Water Supply: the New York City Strategy. Washington, D.C.: The National Academies Press.

National Resources Defense Council (NRDC). 1999. Bottled water: Pure drink or pure hype? Executive Summary, Chapter 2. Available on-line at *http://www.nrdc.org/water/drinking/bw/ exesum.asp*. Accessed February 6, 2005.

Nightingale, S. D., L. T. Byrd, P. M. Southern, J. D. Jockusch, S. X. Cal, and B. A. Wynne. 1992. Mycobacterium avium-intracellulare complex bacteremia in human immunodeficiency virus positive patients. Journal of Infectious Disease 165:1082–1085.

Older Americans. 2004. Key indicators of well-being. Federal Interagency Forum on Aging-Related Statistics. Available on-line at *http://www.agingstats.gov/chartbook2004/population.html*. Accessed February 6, 2005.

Parsons, S., R. Stuetz, B. Jefferson, and M. Edwards. 2004. Corrosion control in water distribution systems: One of the grand engineering challenges for the 21st century. Water Science and Technology 49(2):1–8.

Payment, P. L., E. Franco, L. Richardson, and J. Siemiatycki. 1991. Gastrointestinal health effects associated with the consumption of drinking water produced by point-of-use domestic reverse-osmosis filtration units. Appl. Environ. Microbiol. 57:945–948.

Payment, P. L., L. Richardson, J. Siemiatycki, R. Dewar, M. Edwards, and E. Franco. 1991. A randomized trial to evaluate the risk of gastrointestinal disease due to consumption of drinking water meeting current microbiological standards. American Journal of Public Health 81:703–708.

Payment, P. L., J. Siemiatycki, L. Richardson, G. Renaud, E. Franco, and M. Prevost. 1997. A prospective epidemiological study of gastrointestinal health effects due to the consumption of drinking water. International Journal of Environmental Health Research 7:5–31.

Payment, P., and P. R. Hunter. 2001. Endemic and epidemic infectious intestinal disease and its relationship to drinking water. Pp. 62-88 In Water Quality: Guidelines, Standards and

Health: Assessment of Risk and Risk Management for Water-Related Infectious Disease. L. Fewtrell and J. Bartram (eds.). London: IWA Publishing.

Pierson, G., K. Martel, A. Hill, G. Burlingame, and A. Godfree. 2001. Methods to prevent microbiological contamination associated with main rehabilitation and replacement. Denver, CO: AWWARF.

Reasoner, D. J., J. C. Blannon, and E. E. Geldreich. 1987. Microbiological characteristics of third-faucet point-of-use devices. J. Amer. Water Works Assoc. 79(10):60–66.

Rogers. M. R., B. J. Backstone, A. L. Reyers, and T. C. Covert. 1999. Colonisation of point-of-use water filters by silver resistant non-tuberculous mycobacteria. J. Clin. Pathol. 52(8):629.

Rogers, J., A. B. Dowsett, P. J. Dennis, J. V. Lee, and C. W. Keevil. 1994. Influence of materials on biofilm formation and growth of *Legionella pneumophila* in potable water systems. Appl. Environ. Microbiol. 60:1842–1851.

Rose, C. S., J. W. Martyny, L. S. Newman, D. K. Milton, T. E. King ,Jr., J. L. Beebe, J. B. McCammon, R. E. Hoffman, and K. Kreiss. 1998. Lifeguard lung: endemic granulomatous pneumonitis in an indoor swimming pool. American Journal of Public Health 88(12):1795–1800.

Sandor, J., I. Kiss, O. Farkas, and I. Ember. 2001. Association between gastric cancer mortality and nitrate content of drinking water: ecological study on small area inequalities. European Journal of Epidemiology 17(5):443–447.

Schoenen, D. 1986. Microbial growth due to materials used in drinking water systems. In: Biotechnology, Vol. 8. H. J. Rehm and G. Reed (eds.). Weinheim: VCH Verlagsgesellschaft.

Smith, C. D., and G. Burlingame. 1994. Microbial problems in treated water storage tanks. Proceedings of the 1994 Annual AWWA Conference. Denver, CO: AWWA.

Spinks, A. T., R. H. Dunstan, P. Coombes, and G. Kuczera. 2003. Thermal destruction analyses of water related pathogens at domestic hot water system temperatures. The Institution of Engineers. 28th International Hydrology and Water Resources Symposium.

Stout, J. E., V. L. Yu, and M. G. Best. 1985. Ecology of *Legionella pneumophila* within water distribution systems. Appl. Environ. Microbiol. 49:221–228.

Thofern, E., D. Schoenen, and G. J. Tuschewitzki. 1987. Microbial surface colonization and disinfection problems. Off Gesundh.-wes. 49:Suppl:14-20.

Tobin, R. S., D. K. Smith, and J. A. Lindsay. 1981. Effects of activated carbon and bacteriostatic filters on microbiological quality of drinking water. Appl. Environ. Microbiol. 41(3):646–651.

University of Southern California (USC). 2002. Prevalence of cross connections in household plumbing systems. www.usc.edu/dept/fcchr/epa/hhcc.report.pdf. Los Angeles, CA: USC Foundation for Cross-Connection Control and Hydraulic Research.

Van der Leeden, F., F. L. Troise, and D. K. Todd. 1990. Water quality. Pp. 417-493 In: The Water Encyclopedia, Second Edition. Michigan: Lewis Publishers.

van der Wende, E., and W. G. Characklis. 1990. Biofilms in potable water distribution systems. Chapter 12 In: Drinking water microbiology. G. A. McFeters (ed.). New York: Springer-Verlag.

van der Wende, E., W. G. Characklis, and D. B. Smith. 1989. Biofilms and bacterial drinking water quality. Water Research 23:1313.

Vikesland, P., K. Ozekin, and R. L. Valentine. 2001. Monochloramine decay in model and distribution system water. Water Research 35(7):1766–1776.

von Reyn, C. F., J. N. Maslow, T. S. Barber, J. O. Falkinham, III, and R. D. Arbeit. 1994. Persistent colonisation of potable water as a source of *Mycobacterium avium* infection in AIDS. Lancet 343:1137–1141.

von Reyn, C. F., R. D. Waddell, T. Eaton, R. D. Arbeit, J. N. Maslow, T. W. Barber, R. J. Brindle, C. F. Gilks, J. Lumio, J. Lahdevirta, A. Ranki, D. Dawson, and J. O. Falkinham, III. 1993. Isolation of *Mycobacterium avium* complex from water in the United States, Finland, Zaire, and Kenya. Journal of Clinical Microbiology 31:3227–3230.

Wadowsky, R. M., and R. B. Yee. 1983. Satellite growth of *Legionella pneumophila* with an environmental isolate of *Flavobacterium breve*. Appl. Environ. Microbiol. 46:1447–1449.

Wadowsky, R. M., and R. B. Yee. 1985. Effect of non-legionellaceae bacteria on the multiplication of *Legionella pneumophila* in potable water. Appl. Environ. Microbiol. 49:1206–1210.

Water Quality Association (WQA). 2001. National consumer water quality survey fact sheet. April 23, 2001. Lisle, IL: WQA.

WQA. 2003. Heterotrophic bacteria in drinking water from POU & POE devices. Lisle, IL: WQA.

Wolfe, R. L., E. G. Means, M. K. Davis, and S. E. Barrett. 1988. Biological nitrification in covered reservoirs containing chloraminated water. J. Amer. Water Works Assoc. 80(9):109–114.

Wolfe, R. L., N. I. Lieu, G. Izaguirre, and E. G. Means. 1990. Ammonia oxidizing bacteria in a chloraminated distribution system: seasonal occurrence, distribution, and disinfection resistance. Appl. Environ. Microbiol. 56(2):451–462.

Wood, D. J., S. Lingireddy, and P. F. Boulos. 2005. Pressure wave analysis of transient flow in pipe distribution systems. Pasadena, CA: MWH Soft Pub.

Appendix A

Committee Biographical Information

Vernon L. Snoeyink, Chair, is the Ivan Racheff Professor of Environmental Engineering at the Department of Civil and Environmental Engineering, University of Illinois at Urbana-Champaign. Dr. Snoeyink's research interests involve the physical and chemical processes of drinking water purification, including the removal of trace organic compounds by adsorption onto activated carbon and the development of integrated adsorption-low pressure membrane systems for removing particles and trace contaminants from water. His work has also focused on precipitation of solids in water distribution systems and the chemistry of the formation of colored water in corroded iron pipes. He has been a trustee of the American Water Works Association Research Foundation, president of the Association of Environmental Engineering Professors, a member of the editorial advisory board of the Journal of the American Water Works Association, and vice-chair of the Drinking Water Committee of the EPA's Science Advisory Board. Dr. Snoeyink is a member of the National Academy of Engineering and has participated in several NRC committees as either the chair or a member. He has a B.S. in civil engineering, an M.S. in sanitary engineering and a Ph.D. in water resources engineering from the University of Michigan.

Charles N. Haas, Vice-Chair, is the Betz Chair Professor of Environmental Engineering at Drexel University. His areas of research involve microbial and chemical risk assessment, chemical fate and transport, hazardous waste processing and disposal practices, industrial wastewater treatment, and water and wastewater disinfection processes. He is currently conducting research to evaluate the analytical capabilities for monitoring intentional contamination of drinking water. He has co-authored 14 books or major works on water and wastewater treatment and/or microbial risk assessment. He has served on seven NRC committees, including the Committee for Indicators of Waterborne Pathogens and the Committee to Review the New York City Watershed Management Strategy. He is currently a member of the Water Science and Technology Board. He is a fellow of the American Academy of Microbiology, the Society for Risk Analysis, and the American Association for the Advancement of Science. Dr. Haas received a B.S. in biology and an M.S. in environmental engineering from the Illinois Institute of Technology, and a Ph.D. in environmental engineering from the University of Illinois.

Paul F. Boulos is president and COO of MWH Soft. Dr. Boulos is a recognized world-leading expert on water distribution engineering. He oversees worldwide operations and strategic directions from MWH Soft Corporate headquarters with more than 15 years of industry experience, primarily master planning, hydraulic modeling, and regional water quality studies to large IT integration projects. He has written over 100 technical papers and co-authored two books on water distribution systems analysis. He has also served on technical review committees for several municipal drinking water projects throughout the United States and is currently a Consultant to the EPA Science Advisory Board Drinking Water Committee (Stage 2 Disinfection/Disinfectant Byproduct Rule and Long-Term 2 Enhanced Water Treatment Rule). Dr. Boulos received his B.S., M.S., and Ph.D. in civil engineering from the University of Kentucky and his MBA from the Harvard Business School.

Gary A. Burlingame is the administrative scientist of the Philadelphia Water Department's organic chemistry and aquatic biology laboratories at the Bureau of Laboratory Services, where since 1983 he has been involved in a wide range of operational and distribution system activities. He oversees monitoring of drinking water, source water, wastewater, sediment, sludge, and related media for disinfection byproducts, natural organic matter, PCBs, emerging chemicals, VOCs and SOCs, algae, coliform bacteria, and *Giardia* and *Cryptosporidium*. Best known for his contributions to taste and odor control of drinking water and odor control at wastewater treatment plants, his current research addresses the issues of watershed monitoring and distribution system quality control. He has participated in the review of the Total Coliform Rule that is currently under development and in the expert workshop on Exposure Assessment of Contamination of Distribution Systems. He is the current chair of The Unsolicited Proposal Review Committee for AWWA Research Foundation. Mr. Burlingame received his B.S. and M.S. degrees in environmental science from Drexel University.

Anne K. Camper is a professor of civil engineering, adjunct associate professor of microbiology, and faculty member of the Center for Biofilm Engineering at Montana State University. She is also Associate Dean for Research and Graduate Studies in the College of Engineering. Her primary research interests are in biofilm formation in low nutrient environments, including microbial physiology and ecology, as well as the persistence of pathogenic bacteria in biofilms. Application areas are biological treatment of drinking water and microbial regrowth in drinking water distribution systems. She recently participated in the EPA workshop on Exposure Assessment of Pathogens and Toxic Chemicals in Drinking Water Distribution Systems, the outcomes of which are to be coupled with revisions to the Total Coliform Rule. She is presently on the editorial boards of both Microbial Ecology and Biofilms. Dr. Camper received her B.S. and M.S. in microbiology and her Ph.D. in civil and environmental engineering, all from Montana State University.

Robert M. Clark is an environmental and engineering and public health consultant. He is a consultant to Shaw Environmental and Infrastructure (SE&I) working on homeland security issues and to the University of Cincinnati working on risk assessment methodology for water system vulnerability. He spent over 40 years in government, first for the U.S. Public Health Service Commissioned Corps, and then for EPA where he directed the Water Supply and Water Resources Division for 14 years. Among other things, his research interests have focused on modeling water quality in drinking water distribution systems, including understanding the many factors that can cause the quality of distribution system water to deteriorate such as the chemical and biological quality of source water, the effectiveness and efficiency of treatment processes, the adequacy of the treatment facility and storage facilities, distribution system age and design, the maintenance of the distribution system, and the quality of treated water. He received the 2004 Lifetime Achievement Award from the American Society of Civil Engineers' Environmental & Water Resources Institute. Dr. Clark holds a B.S. in civil engineering from Oregon State University, a B.S. in mathematics from Portland State University, a M.S. in mathematics from Xavier University, an M.S. in water resources and environmental planning from Cornell University, and a Ph.D. in environmental engineering from the University of Cincinnati.

Marc A. Edwards is the Charles Lunsford Professor of Civil Engineering at Virginia Polytechnic and State University. Prior to joining the faculty at Virginia Tech, he was an assistant professor at the University of Colorado in Boulder and a senior engineer with Montgomery Engineers. Dr. Edwards is the current president of the Association of Environmental and Engineering Science Professors. His research interests are internal corrosion processes in home plumbing, water treatment, scaling, arsenic removal, and applied aquatic chemistry. The White House honored him in 1996 with a National Science Foundation Presidential Faculty Fellowship, an award that was given to only 20 professors annually. In 2003 he was awarded the Walter Huber Research Prize from the American Society of Civil Engineers. Dr. Edwards received a B.S. in biophysics from the State University of New York and an M.S. and a Ph.D. in environmental engineering from the University of Washington.

Mark W. LeChevallier is chief scientist for innovation and technology at the American Water Corporate Center in Voorhees, NJ, which owns and operates numerous drinking water utilities throughout the United States. His research involves a wide area of issues in drinking water distribution systems, including bacterial regrowth, disinfection of biofilms, corrosion, bacterial nutrients, AOC measurement techniques, biological treatment, Mycobacterium, microbial recovery and identification, the impact of pressure transients on water quality, and detection, treatment, and survival of *Giardia* and *Cryptosporidium*. He recently participated in an expert workshop on Exposure Assessment of Contamination of Distribution Systems, which resulted in several white papers that formed the basis for the current study. Dr. LeChevallier currently serves as the Chair of the AWWA Microbial/Disinfection By-Product Technical Action Workgroup and is a trustee of the AWWA Water Science and Research Division. He received his B.S. and M.S. degrees in microbiology from Oregon State University and his Ph.D. in microbiology from Montana State University.

L. D. McMullen is the CEO and general manager of the Des Moines, IA, Water Works where, among other accomplishments, he supervised one of the Upper Midwest's largest "design and build" concept water plant projects. Dr. McMullen served two terms as Chair of the National Drinking Water Advisory council of EPA and on the Science Advisory Board's Drinking Water Committee. He has served as a diplomat for water supply/wastewater issues for the American Academy of Environmental Engineers, as the Water Industry Delegation Leader to China for the Citizen Ambassador Program, and was an executive committee member of the Board of Trustees of the American Water

Works Association Research Foundation. In 1994 Dr. McMullen received the President's Award of the Association of Metropolitan Water Agencies. He received his B.S. in civil engineering and an M.S. and Ph.D. in environmental engineering from the University of Iowa.

Christine L. Moe is an associate professor of infectious diseases in the Department of International Health at the Rollins School of Public Health at Emory University. Previously she was an assistant professor in the Department of Epidemiology, University of North Carolina, Chapel Hill. She received her Ph.D. in environmental sciences and engineering from the University of North Carolina and has done extensive laboratory and field research on waterborne transmission of infectious agents and diagnosis and epidemiology of enteric virus infections. She is a member of the WSTB and also served as a member for the NRC Committee on Evaluation of the Viability of Augmenting Potable Water Supplied with Reclaimed Water and the Committee on Watershed Management for New York City.

Eva C. Nieminski is an environmental research engineer at the Utah Department of Environmental Quality Division of Drinking Water, where she provides technical assistance to 50 water treatment plants in Utah. She is also an adjunct associate professor in the department of civil and environmental engineering at Utah State University. Her research focuses primarily on treatment of drinking water, including filtration, UV disinfection, the application of surrogate measures to improve treatment plant performance, *Giardia* and *Cryptosporidium* removal via conventional treatment and direct filtration, ozone pilot studies, disinfection byproduct studies, and a TOC study in surface water treatment plants. She serves as a trustee of the AWWA Water Quality Technology Division. In the regulatory arena, she has represented the Association of State Drinking Water Administrators on negotiated rule making for the Disinfection Byproducts and Enhanced Surface Water Treatment Rule, as well as ECOS for Stage II Disinfection Byproducts and Long-Term Enhanced Surface Water Treatment Rule. Dr. Nieminski received her M.S. in civil and environmental engineering from Warsaw Technical University in Poland, her M.S. in environmental health engineering from the University of Notre Dame, and her Ph.D. in civil and environmental engineering from Utah State University.

Charlotte D. Smith is president of Charlotte Smith & Associates, Inc, which provides consulting services to drinking water utilities nationwide. Before establishing Charlotte Smith & Associates, Inc., Ms. Smith worked with the New York City Department of Environmental Protection's Drinking Water Quality Division, and she was director of Water Quality for United Water Resources (formerly General Waterworks Corp.), which operated 35 drinking water utilities in 15 states. Ms. Smith's expertise with respect to distribution systems has spanned from understanding the effect of treatment plant practices and chemical and biological stability on distribution system water quality, to disinfectant residual studies, corrosion studies, nitrification control, and distribution system tracer studies. She is a member of the American Water Works Association Distribution System Water Quality Committee (immediate past chair) and Microbial/Disinfection By-Product Technical Advisory Group. Ms. Smith led the development of a Distribution System Self-Assessment Workbook for drinking water utilities. She holds a B.S. in microbiology from the University of Michigan and an M.S. in community health from the City University of New York.

David P. Spath is chief of the Division of Drinking Water and Environmental Management at the California Department of Health Services, where he has worked since 1972. He is currently responsible for overseeing California's Public Water System Regulatory Program, its Medical Waste Regulatory Program, and the state's Nuclear Emergency Response Program. He is chair of the National Drinking Water Advisory Council and also serves on the California Recycled Water Task Force. He is past president of the Association of State Drinking Water Administrators and served on a steering committee for EPA's environmental technology verification program related to small water systems. He was a member of the recently concluded NRC study on Water System Security Research. Dr. Spath received his B.S. in civil engineering from Tufts University and his M.S. and Ph.D. in civil and environmental engineering from the University of Cincinnati.

Gary A. Toranzos is an associate professor in the department of biology at the University of Puerto Rico. His current research involves analyzing survival of genetically engineered microorganisms (GEMs) in the environment; using molecular techniques such as gene probes and polymerase chain reaction to detect pathogens and GEMs; analyzing transfer of genetic material under environmental conditions; exploring the potential for using indigenous or genetically manipulated organisms for biotransformations of xenobiotics; and developing and testing indicators for fecal contamination in water. He was a fellow of the Latin American Professorship Program of the American Society for Microbiology in 1991, an elected fellow of the American Association for the Advancement of Science in 1995, and an elected member of the EPA's Science Advisory Board from 1997 to 2004. Dr. Toranzos has a B.S. in microbiology and chemistry, an M.S. in environmental microbiology, and a Ph.D. in environmental virology and food science—all from the University of Arizona.

Richard L. Valentine is a professor in the civil and environmental engineering department at the University of Iowa. He is also a member of the Center of Health Effects of Environmental Contamination. Dr. Valentine's current research interests are in the general areas of environmental chemistry and physical and chemical processes in natural and engineered systems, especially water and wastewater treatment process design and modeling; environmental chemistry/reaction kinetics; processes to remove trace contamination from water; and fate and transformation of hazardous chemicals. His current research related to distribution system issues includes the chemistry of disinfectants and radium and radon in drinking water; mineral dissolution processes; the use of metal oxides as adsorbents in drinking water treatment; and role of the pipe-water interface in the determination of drinking water quality. He received B.S. degrees in chemical engineering and chemistry and an M.S. in chemical engineering from the University of Michigan; and an M.S. and a Ph.D. in civil and environmental engineering from the University of California at Berkeley.